Praise for *Signs of Life*:

'A fever dre[illegible] mcha[illegible] [illegible] grand tour of six years on two wheels ... I [illegible] a couple of days to it and on emerging from its pages the world had become richer and more marvellous ... *Signs of Life* deserves to become a classic of the genre'

Gavin Francis, author of *Adventures in Human Being*

'Cramming 53,568 miles into 416 pages is a feat as great as pedalling the world. I really enjoyed it. I was with him over every bump. A helter-skelter, surgical circumnavigation. It's unusual to cover so much ground and yet also to manage to be so anatomical about places and people. It's incredibly vivid. The medical lens is very funny, humane and revealing. A wildly energetic adventure to unreachable lands'

Nicholas Crane

'A clever and entertaining book. Fabes has a gift for describing landscapes ... but he isn't afraid to share those lurid medical anecdotes we all crave'

Jonathan Drummond, *TLS*

'Witty and wild, intrepid and inspirational, the book chronicles two parallel journeys: Fabes's physical cycling tour of many countries and his look at health across the globe ... an entertaining and epic chronicle of a journey of extremes'

Tony Miksanek, *Booklist*

'A charming, human story of resilience, adversity and compassion, all told with a good dose of dry British humour'

Lois Pryce, author of *Revolutionary Ride*

'In Stephen Fabes's captivating odyssey we probe not just the dizzying expanses of our planet's geography but also the exquisite breadth of the human spirit. In an age of weary cynicism, it's reassuring to know that there are still signs of life out there'

Danielle Ofri, MD, PhD, author of *When We Do Harm: A Doctor Confronts Medical Error*, and Editor-in-Chief of *Bellevue Literary Review*

'*Signs of Life* is a plucky memoir ... Fabes is a winning storyteller. During his journey, Fabes battled injuries, infestations and dengue fever. Yet his travelogue retains its bighearted humour'

Kevin Canfield, *Star Tribune*

'Stephen has managed to beautifully combine travel, medicine, life and loss all in this epic tale of adventure. It made me smile, laugh, want and cry. Bloody brilliant'

Matt Morgan, author of *Critical*

'A hilarious and moving memoir of a cycling medic's life on the road'

Xand van Tulleken

'*Signs of Life* is the kind of book we need right now. It is heart-warming. It is hopeful. It shows us that despite all the guns and guerrillas the world is also full of people who press presents and hospitality on travellers. It is also a very readable travelogue, a thoughtful one'

Lesley Mason, The Bookbag

‘I absolutely loved this book – a six-year, wheel-spinning, page-turning pedal around the world. Full of rich, vivid descriptions of people and landscapes, interwoven with wisdom, compassion, humanity and a playful Panglossian humour … he allows us to see how everything is connected. If Stephen is half as talented a doctor as he is a writer, then lucky patients I say. A worthy addition to the cycle-touring canon’

Mike Carter, author of *One Man and His Bike*

‘The humour, humility and self-awareness reminded me at times of Eric Newby. Local history, politics and anecdote are woven together, introducing the reader to a world most will never encounter, and I for one felt richer for the experience … It’s hard to believe this is Stephen’s first book; I sincerely hope it’s not his last’

Sam Jones, *Cycle* magazine

‘I was smitten with this book. It was the world writ large, framed with self-deprecating, glorious humour. If, like me, you have a hankering to lose yourself in the wilderness, to race semi-naked into uncertainty, beating your soft city chest and howling at the moon, this book can let you imagine it. As Fabes put it, his rusty bicycle gave him a backstage pass to the world. By writing the book so marvellously, he gave us one as well’

Michelle Johnston, *Life in the Fast Lane* blog

SIGNS OF LIFE

To the Ends of the Earth with a Doctor

STEPHEN FABES

PURSUIT

This paperback edition published in 2021

First published in Great Britain in 2020 by
Profile Books Ltd
29 Cloth Fair
London
EC1A 7JQ

www.profilebooks.com

The author and publisher would like to thank Naomi Rowsell (2, 3, 8), Alistair Hill (18) and Sam Lovell (31) for the use of their photographs. All other photographs by the author.

1 3 5 7 9 10 8 6 4 2

Typeset in Sabon by MacGuru Ltd
Printed and bound in Great Britain by
CPI Group (UK) Ltd, Croydon, CR0 4YY

A CIP catalogue record for this book is available
from the British Library.

ISBN 978 1 78816 122 0
eISBN 978 1 78283 477 9

For Mum, of course

Contents

Each of us is a biography, a story. Each of us is a singular narrative, which is constructed, continually, unconsciously, by, through, and in us – through our perceptions, our feelings, our thoughts, our actions; and, not least, our discourse, our spoken narrations. Biologically, physiologically, we are not so different from each other; historically, as narratives – we are each of us unique.

Oliver Sacks, *The Man Who Mistook His Wife for a Hat*

Arctic Ocean
Deadhorse
Chicken
Fairbanks
Whitehorse
CANADA
Vancouver
Portland
Ottawa
USA
San Francisco
Washington DC
Los Angeles
Tijuana
Baja California
Atlantic Ocean
MEXICO
Mexico City
Oaxaca
Cartegena
Panama City
Bogota
Pacific Ocean
Quito
PERU
BRAZIL
Lima
La Paz
Salar de Uyuni
Paso de Sico
Paso Internacional Los Libertadores
Santiago
CHILE
Buenos Aires
Puerto Montt
Ushuaia
London
Calais
FRANCE
Paris
Geneva
Nice
Southern Ocean
ANTARCTICA

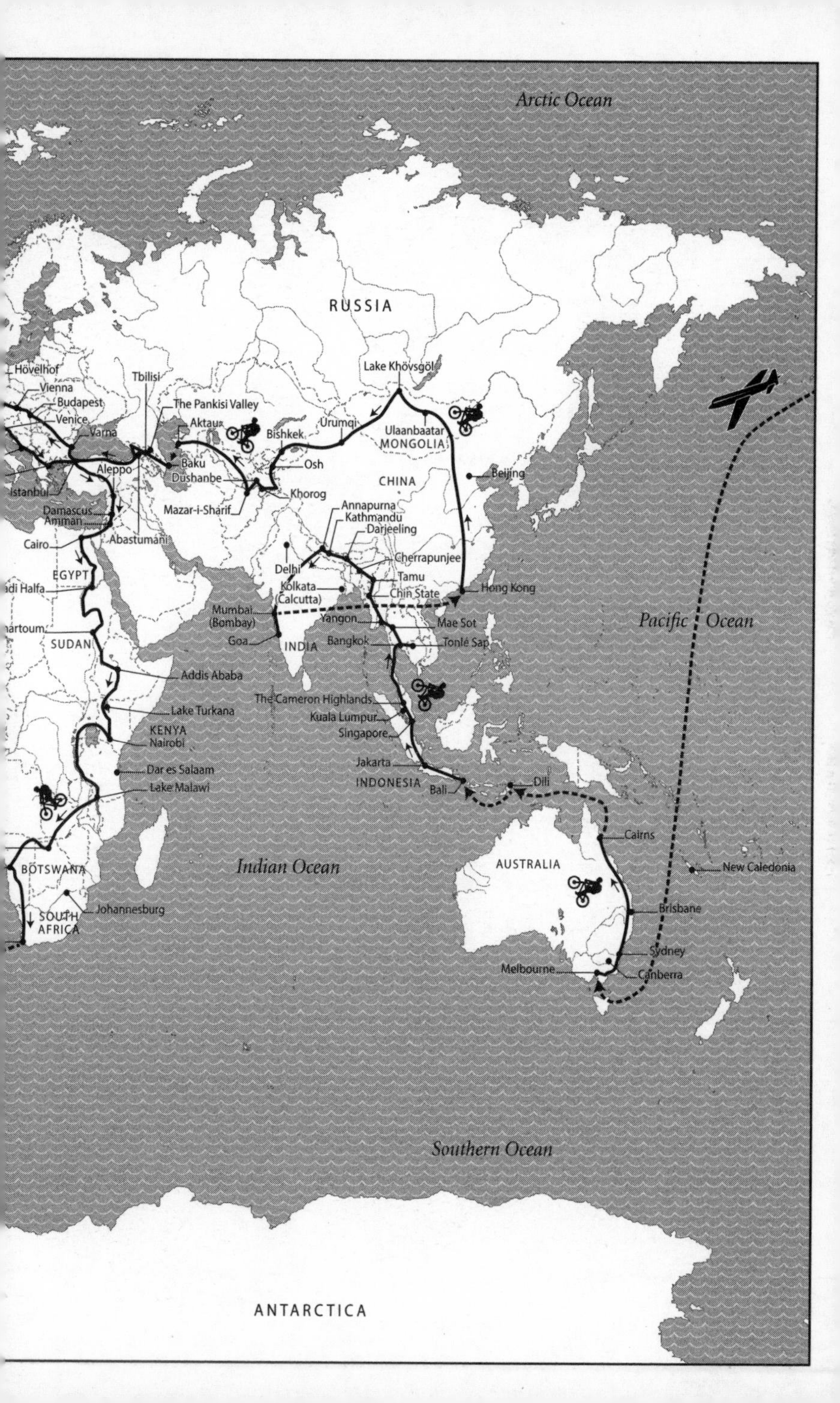

Arctic Ocean
RUSSIA
Hövelhof
Vienna
Budapest
Venice
Varna
Tbilisi
The Pankisi Valley
Aktau
Baku
Dushanbe
Mazar-i-Sharif
Bishkek
Urumqi
Osh
Khorog
Lake Khövsgöl
Ulaanbaatar
MONGOLIA
CHINA
Beijing
Istanbul
Aleppo
Damascus
Amman
Abastumani
Cairo
EGYPT
adi Halfa
hartoum
SUDAN
Addis Ababa
Lake Turkana
KENYA
Nairobi
Dar es Salaam
Lake Malawi
BOTSWANA
Johannesburg
SOUTH AFRICA
Annapurna
Kathmandu
Darjeeling
Delhi
Cherrapunjee
Tamu
Chin State
Kolkata (Calcutta)
Mumbai (Bombay)
Goa
INDIA
Yangon
Mae Sot
Bangkok
Tonlé Sap
Hong Kong
The Cameron Highlands
Kuala Lumpur
Singapore
Jakarta
INDONESIA
Bali
Dili
Pacific Ocean
Indian Ocean
AUSTRALIA
Cairns
Brisbane
Sydney
Canberra
Melbourne
New Caledonia
Southern Ocean
ANTARCTICA

Prologue

There is a moment I like to think of as a landmark in the six years I spent cycling around the world, a moment that revealed the absurd totality of what I had done. It came by accident, in Kent. My bike computer had clocked over 53,000 miles and I was finally closing in on my friends and family who would gather tomorrow in London to welcome me home. After struggling into a headwind all morning – a situation, I'd decided, that called for doughnuts – I locked my bike to a lamp post in Canterbury, walked into Sainsbury's and began pushing a shopping trolley down the aisle.

And then, quite thoughtlessly, my right hand rotated around the trolley's handle. I paused by the chicken Kievs, peered down, and realised. I was trying to change gear.

Thinking back, I wonder if leaving home was a kind of instinct too. But where does it come from, the itch to stray, to go across, around and beyond, to probe whatever's over *there*? For me at least, it began with another journey. I was nineteen years old, cycling across the plains of Patagonia with my younger brother, Ronan.

We were planning to ride the length of Chile, from bottom to top. It was my idea. I'd decided in my boxy bedroom at home in Oxford as I explored the globe on my window sill. My brother and I were restless teenagers, yearning to be at large in the world. Just the name, *Chile*, conjured somewhere half-real, the names of the towns too – Osorno, Temuco, Curico – urged a closer look. In my granddad's weathered atlas – a massive thing, even in an adult's hands – Chile clambered improbably up the map, crinkled with mountains and streaked in different colours. In the map's key, Chile was transformed into a hotchpotch of worlds: wide plains, chains of volcanoes, monkey puzzle forests, nibbled coastlines of shiny white sand. It stretched north into the Atacama Desert too, where the odd dots of towns stopped and tantalising space began. The view beyond my bedroom window was painfully suburban: a stale row of pebble-dashed, semi-detached houses and carefully parked Ford Fiestas. Nothing interesting ever happened on my street.

Nothing, until a few weeks before my brother and I left for South America, when a conspicuous stranger appeared. I watched him through the window drifting up and down our road on a touring bike with drop-down bars and a couple of grey panniers on the rack. He was scanning house numbers, and, noticing ours, pedalled over. I opened the door to an outrageous beard. The man stood gangly and grinning in a retro cycling cap and weather-worn vest. His shoulders were lean and toasted by the sun. He explained that he'd read about our plans to cycle through South America in a local newspaper – where he'd discovered our address – and had come to see how our preparations were going. Perhaps, he said, he could offer some advice. He'd cycled the length of the Americas himself not too long ago, from Patagonia to Alaska, three years of wandering the back roads, living cheap and camping wild. He brought to mind *Forrest Gump*, of course, as any grizzled, beardy and relentless traveller might. I'd loved how Forrest runs back and forth across America without any particular design or destination, and so it was Forrest that my brother and I named this stranger at the door.

My mum invited Forrest inside, seeking peace of mind. Naturally, she had concerns – we had no other siblings and no father at home – and it can't have helped that she taught in a secondary school and knew all about juvenile delusions of invulnerability. She must have seen this in us, even recognised it, perhaps, as a blessing mixed with a curse. We were *that* age after all, the age at which you're most likely to surf a moving car or attempt to 'drink' vodka through your eye because Gary says it gets you pissed faster.

Of course, Forrest didn't know that my mum was bracing for catastrophe as he sank himself into an armchair and stretched out sinewy legs. Spurred by my questions, he began to tell stories from those years on a bicycle. Most were horrible. One morning, camped out in the Colombian jungle, he'd woken to find his legs covered in leeches. An Andean storm had overwhelmed his tent with snow. He spread out a map on the living-room floor and the roads came alive: here, he'd found a puma, beheaded, strung up, rank, to a fencepost. My mum cringed, which delighted me at the time. She asked him if he wore a helmet.

'Nah! One of them trucks in Argentina hits you and a helmet won't matter. You'll be smash—' he stalled, noting my mum's alarm.

'Yeah sure. You guys should wear helmets.'

And he looked at me sideways, with all the mischief of a wink.

This was the first time I'd met a travelling biker and I was beguiled. Forrest, though, was hardly alone. Thousands had cycled the length and breadth of continents by then, and many others had looped the world. A few extremists had set out on unicycles, or detoured through Siberia, or had even jumped into pedalos for the watery bits. And all for what? Could that misty premise of 'adventure' be enough? What even was that?

As a teenager, hungry for answers and adventures, I went looking in books. At first, I journeyed with the explorers, scaling lung-crushing summits, slogging across frozen wastes, sometimes plunging into gory fights for survival. These tales were steeped in machismo; their heroes wrote of comradery, and of daring. But in time, I drifted away from these solid, formidable men. Later, I found the likes of Dervla Murphy, who roamed from Ireland to India on her bicycle, and pedestrians like Laurie Lee, who plodded from the Cotswolds to Spain in the 1930s, and the bumbling naturalist Redmond O'Hanlon, who wriggled through jungles, soaked and bitten and still laughing. These wandering writers didn't count themselves as explorers, and I found them easier to love, perhaps because they were more outward-looking, not obsessively self-reflective, or perhaps because the act of writing sustained their curiosity and charged their adventures. There was still high drama – in *Full Tilt*, Dervla is coolly blasting wolves with her revolver by page 8 – but I admired how they travelled slow and thoughtfully, questioning as they went. Often, explorers were adversarial; they approached the world as a challenge to be overcome, and wrote in terms of rule and dominion over nature. Travel writers, on the other hand, suggested that there were better reasons to go on adventures beyond overcoming a physical test, or the dubious inducements of fame and conquest. They dug beneath the myths of a place, and fell in with local people. They sought insight and awe and a fresh perspective on the world – these were higher ambitions, surely, than a simple peak or traverse. And they were meanderers, always, who detoured on a whim and resisted straight lines.

The journey that my brother and I eventually made from one end of Chile to the other was equal parts disaster and revelation. Disaster for all the predictable reasons: we were young, starry-eyed and broke. We'd sponged most of our cycling kit from companies sympathetic to our age and our outsized dreams (grandly, we called this sponsorship), but its quality matched what we'd paid for it, which was nothing. Within the first week, my aluminium pannier rack buckled into my whirling spokes and I was tossed from my bike like a rodeo cowboy.

We improvised to get going again, tethering kit to the frame of my bike, and for the next few weeks I tracked Ronan across Patagonia – a widespread monotony of rocks and shivering scrub. A squally wind picked up and we were socked this way and that, our tyre tracks winding together, leaving something like a double helix in the rock dust of the trail, a vague impression of life where there was little in any direction to see. Gale force wind was not a weather event out here, more a mood of the place, a habitual violence that purged and levelled the land.

Perhaps it was the wind that made Patagonia feel so wild and dicey. Joan Didion, writing about other gales, the dry Santa Ana 'devil-winds' buffeting Los Angeles, believed that they heightened the city's mood of impermanence and unreliability. 'The wind', she wrote, in *Slouching Towards Bethlehem*, 'shows us how close to the edge we are.' I felt near some kind of edge in Patagonia too; there was something in the howling severity I admired. I'd never felt so tiny and wind-tossed, so blissfully vulnerable before.

My brother and I never stayed anywhere long, sleeping beside roads wherever we pleased, and the arbitrariness of our campsites was pure liberation. We chose routes with impulsive pleasure and we ate like escaped prisoners of war. Mountains rose and fell and there was an abiding sensation of being dwarfed by the landscape, as if we were animals filmed from above in a wildlife documentary. It all feels a long time ago now, and in my memory roads flow together as one, but from time to time I dust off photos because they tell a better story than I can. We look happy and self-possessed. Fuck! – we're glowing.

Regardless of how the terrain behaves, travelling long distances by bicycle can be an undulating task, awesome in one moment, deathly boring the next. I enjoy this turbulence, this hovering sense that your luck might turn at any minute – sometimes all it takes is a gap in the clouds and a big, heart-thumping view, or a stranger cheering encouragement and handing you a snack from the window of a car. On the back roads, I felt nurtured by wildness, never at war, and, often, profoundly at peace.

Chileans tended to treat us with great warmth as we travelled. Our age must have helped, but then so did our means. Travelling so gradually, so blatantly, the roving cyclist is exposed to the kindliness of others, and soon you sense it seeping from every pore of the planet. My brother and I wobbled into countless Andean villages where we were quickly lent places to sleep, in homes and schools and police stations and army barracks. We were tucked under woollen blankets, our bellies full of hot stew. In these small favours, and relentless good fortune, my bike – it seemed to me – was a backstage pass to the world.

Five months after setting off, amid the quiet sands of the Atacama, I was reminded how far we'd come from the baying winds of the south. Studying my legs, and their strange new muscles, had the same effect. Riding my bicycle a long way had proved surprisingly simple. What if I kept going? I'd reach the cloud forests of Colombia in a few months' time, the wide beaches of Oaxaca not long after that. I could explore the world by degrees, and I could make it to Alaska if I wanted to, a compelling wildness, patrolled by grizzly bears. Possibilities sprawled.

*

Instead I flew home, and life, of course, took over. I began medical school in Liverpool that autumn and I was flung into five years of study, five years to wonder at the human machine and the dizzying ways in which a life could be undone. After graduating from lecture halls, from the formaldehyde stench of the dissection room, the towers of textbooks and masses of revision cards, I worked for two years in a hospital in Merseyside as a junior doctor. Living with other junior doctors was an adventure of peaks and troughs too, all of us hoping to make sense of our role, everyone out of their depth and desperate not to add a big screw-up to the inevitable string of smaller ones. In those first years, hardly a week went by without a first time: the first time I'd given a diagnosis with life-changing consequences; the first time I'd shocked a patient's heart; the first time this had failed to restore a pulse; the first time I'd certified a death or explained its inevitability to relatives; the first time I'd helped remove a man's foreskin from his flies.

I wish that I could say I became a doctor through an innate compassion for humankind, or something equally tender, but, to begin with at least, that was, sadly, not my instinct. I had applied to medical school because I enjoyed science and my teachers cautioned that it was tough to get in. It felt like a dare, and I flung myself towards the challenge itself, unthinking. It terrifies me now that I hadn't seriously considered the greater challenges on the far side of that medical school interview, the challenges that decided the lives and deaths of human beings.

Luckily, then, I've never seriously doubted my decision to become a doctor. The hours can be treacherously long and the work can be traumatic, but mostly it is a generous job, and high-octane at times, full of problems to solve and questions to answer. Even before you're awed by the finer points of human physiology, a job in which strangers candidly share the landscape of their lives with you – their wishes and fears, their figments and secrets, their *stories* – is never dull for very long. Storytelling lies at the heart of my job: sometimes my patients are supreme storytellers, but even when they're not, you might divine the story from what's not said or half said, or even from the test results.

After two years of foundation training in Merseyside, I landed a job as a Core Medical Trainee – a step up – in a prestigious place. St Thomas' Hospital rises ivory-white over the Thames and confronts the Houses of Parliament. It is a large teaching hospital with a venerable history, copious experts and a legacy of medical pioneers. A shift in its emergency department can reveal what a mixed-up city London can be, from dizzy, hypotensive Lords tottering over from Westminster, to the heart-racing ravers of Vauxhall, fresh from ravenous, chemsex parties lasting three days. You're never quite sure what you'll hear next.

For a couple of years, I rotated happily through specialties at Guy's and St Thomas', inserting central lines and chest drains on intensive care, dashing between crash calls, scurrying around the Renal Transplant Unit prepping fathers and sons, sisters and brothers, for surgery. But at the same time I began to wonder whether I'd fall into line and march steadily up the ranks towards a consultant post. If loving my work was a privilege at first, it felt inconvenient now, obstructive even, however ungrateful that may sound. When I was asked to choose a specialty, I sensed a narrowing not just of my field of practice, but in my life and opportunities too. I was at the dog-end of my twenties, and as a new decade loomed, there was a sense of time plunging away, a fear that I'd wake up one morning with erectile dysfunction, or worse, a passion for quilting and jigsaws.

It was not so much a longing to escape, though, not a push. I recall feeling pulled towards something instead: but to what? I couldn't label the feeling then, but I have since discovered a German word, *sehnsucht*, which comes close. *Sehnsucht* suggests a rapturous pining for *more*, but it's a *more* that can't be easily described or defined. I needed more of almost everything, more space, more time, more risk: more adventure too, because I sensed it would make me in some way, rewire me, revise my sense of the world.

I never thought for a moment that I'd give up being a doctor for good. I relished the ceaseless learning, the teamwork and the frank and privileged view of worlds I otherwise wouldn't see, worlds of patients and the idiosyncrasies of their lives. But when I slipped back into memories of riding through Chile, I couldn't shake the feeling that an opportunity was slipping away. The now-or-never moment arrived as other junior doctors were applying for specialty training posts. I gritted my teeth and hung tight.

It's tough to stop working as a doctor for almost any reason, but especially for something as casual as travel, and not feel pinched by guilt. You may meet a small but clear undercurrent of disapproval too, from those colleagues who would never prioritise anything over medicine, including, in some cases, family and the chance of a wider life. When I thought of leaving my job, I did wonder whether selfishness was trumping my sense of wanting to serve and be useful. There are ways to bury this feeling. The best, I've found, is to remind yourself that you're still young, that you'll be hard at work in the NHS for decades to come. A *Times* mini-atlas helps too. I pored over it in the evenings and dreamt of adventures during the spare moments of the day: in the shower, on the bus, while delivering the chest compressions of cardiopulmonary resuscitation.

I was diverted as I studied for my postgraduate exams too. Hunched over Tortora's *Principles of Anatomy and Physiology*, I began to see homology between the earth and the body in the language of medicine. The islets of Langerhans are clusters of cells in the pancreas, the brain has aqueducts, and the pelvis has an inlet. Even when the language itself isn't evocative of place, the metaphors are clear. An artery can bifurcate like a confluence of rivers; sinuses could be caves, or lakes. A road map of, say, China, links and meanders like a neuronal network. The skin comes in layers, the dermis, the epidermis and the subcutaneous tissue, just as the atmosphere has the troposphere, stratosphere, mesosphere and thermosphere. Or perhaps the three-layered meninges, the linings of the brain, are better suited to that analogy, since both the brain and the planet have hemispheres. We have even mapped the body and the earth in similar ways. Langer's lines are topological lines drawn on a map of the body that correspond to the natural orientation of collagen fibres. They help surgeons to navigate, to know where to place their blade, just as lines of contour or longitude have helped travellers in their wanderings.

A plan was forming, cosmically. An idea sparked in Chile was now a raging chain reaction. A country, or even a continent, wasn't enough this time, and I'd scribbled something far more ambitious in my mini-atlas, a journey spanning the length of each continent. 'Isn't that quite cold?' my mum quibbled, taking in the hopeful line I'd drawn in about three seconds across Antarctica. *Fine.* So I'd ride across *six* continents: Europe first, then Africa, South America, North America, Australia and Asia – I could think of no better way to feel the scale and diversity of the globe. I'd tackle each loosely, knowing only where I'd begin on one pole of the continent, and where I'd finish on the other, without a precise plan of how to pilot the meat in between. Space for serendipity, for risk.

Then there were some hard realities to consider. I was looking at several years without a job or steady income, so I began saving money. At first this meant wedging myself into a flat owned by the NHS. I could barely stand up in my room, the taps dripped and the cold got in. The whole experience was like living in the brig of a North Korean warship. Inevitably, though, life would take a harsher turn once I left home. I could wave my social life goodbye for a start. To stick to my budget – I'd planned on ten dollars per day, tops – I'd have to rough camp, most nights, beside the road. There would be years of porridge and packet noodles, and discomforts that I couldn't predict. At times, I wondered why leaving home felt like such an obvious decision at all. Now, I think I was simply longing for a less certain future. And uncertainty, whether in life or bike rides, is the heart and soul of any journey.

Part One
LONDON TO CAPE TOWN

Bicycling has now lived down the prejudice which, from a medical point of view, existed against it. It is admitted that the idea of rupture being produced by it is simply nonsense.

The Bicyclists' Pocket Book and Diary, 1879

1

Get Him in His Stupid Face

Bike fine-tuned, maps in rainproof cases, panniers packed, tick, tick, tick, and still time for that final, uninvited moment – but one somehow crucial to any big journey – the moment when doubt sets in. No special reason behind it, just a jumble of fears and forebodings and a dim idea that I'd thrown my chips in, staked a comfortable life on a fantasy. The start line wasn't helping either: two friends, Tom and Eddie, were holding out a stretch of red and white barrier tape, the type used to keep people away from hazardous things.

I wheeled my new touring bicycle towards the tape and gave a weak smile to the small crowd that had collected on the forecourt of St Thomas' Hospital to send me off – my mum, foremost, friends, and a few harried-looking doctors and nurses on slim lunch breaks, sandwiches in hand. I was dressed for adventure, but not for style: waterproofed, sensibly helmeted, muffled in base layers and clinging cycling apparel. It was early January, and the weather was nothing special: a dirty-white sky hung over London, and there wasn't enough breeze to spread the cigarette smoke from two grey-faced men with chronic respiratory disease, who stood by the entrance in hospital gowns. Hunched over a sign that said 'This hospital is smoke-free', they were using its metal pole as an ashtray. It was just another daft detail of the NHS but I was in such a wistful mood that it made me feel suddenly sorry to be leaving it all behind.

In all my bookish explorations, I had admired the adventurers who'd set off from home in a casual way: no spectacle, no fuss – a nod, perhaps, to how journeys are full of unknowns. And maybe a bit of humility early doors was a good-luck charm too. But as I scanned the audience now, I could see that times had changed. Millennials – and I'm borderline here – fuck about before they do stuff like this. They launch blogs, throw leaving parties, make a noise on Twitter. If you're having doubts, though, a bit of hype has hidden value – it's much tougher to quit if a hundred people watch you begin.

Lift off! With cheers and applause in my ears, I pedalled through the start line and across the hospital's forecourt, tacking around a few tourists on the end of Westminster Bridge. When six years of cycling loom implausibly ahead, you tend to focus on the next few metres of pavement, and then you photobomb an Asian bride on a wedding shoot because you're not looking where you're going. Perhaps my dubious face survives somewhere, on a mantelpiece or in a photo album, immortalised where Big Ben should have been.

That day was the beginning in one sense, but it was also the end. Two years of preparations and planning were behind me now. Those bottomless to-do lists had finally bottomed out, and crucial questions had been answered and decisions made. (What, for instance, is a more intrepid colour scheme for bicycle panniers, granite/black or lime/moss?*) I'd spent hours lunging around outdoor shops in different brands of padded Lycra shorts wondering if I'd ever get used to feeling like an adult in a nappy. I'd set up a blog that announced my goal, 'cycling six continents', and when I thought hard about it, this struck me as an incredible plan. Sadly, it was only after I'd

* It's granite/black.

cycled a hundred metres to Embankment that it seemed incredible for me in particular.

Besides having a sense of direction that friends considered hilariously lame and an inborn stupidity when it comes to all things mechanical, it was now over a decade since I'd biked the length of Chile with Ronan. I'd managed very little cycling or any serious exercise since working as a doctor and this remained true in the build-up to leaving. Given that cycling around the planet would be tough, I reasoned, why add to the toughness with months of training as well? And so it was taking some effort now to overtake a jogger beside the Thames.

I could blame the 40 kilograms of kit, or the bike itself, which had a hefty steel-frame ('built to last!' according to the guy in the shop, desperate for a sale). But I only had to cast my eyes downwards to see that these were not the most convincing excuses for my pace. If the legs of a seasoned triathlete were at one end of the scale, my pallid, wobbly appendages were at the other. A hollow feeling settled inside me, something hard to nail down. This couldn't be homesickness or loneliness yet – I was not quite at London Bridge – though it was perhaps the anticipation of those things. I was pre-lonely, pre-homesick.

I'd been cycling around the world for twenty minutes or so when a pub called The George appeared on my left. There was a small beer garden, somewhere where I could collect my thoughts, pick at chunky chips, recheck my maps and forget the time. A few friends joined me, and when at last Henry pointed out that it would be getting dark soon, I said goodbye for a second time, wheeled my bike back to the road and began again. My first day as an adventurer ended an hour later, in a guesthouse in Bexleyheath. I drifted off under floral sheets, 14 miles from the start line, a fact that seemed to make my sore arse a particularly fateful detail.

*

I woke up fast, blinking. Through a gap in the curtains, intense, white light was flooding the room. I staggered over to the window, pulled the curtains apart and gaped. Bexley was smothered for as far as I could see. The snow was still falling too, and with each gust of wind, flurries went swirling across the sky.

I'd have cursed my luck if my judgement wasn't so patently to blame. I'd set my departure date with a bold indifference to the seasons, planning simply to leave when I felt ready, which meant attending sufficient Expedition Planning Seminars (four) and cycling expos (three). And here I was, 'ready' during what would be, according to some forecasters, the coldest winter in western Europe for thirty-one years.

I began cramming socks and toothpaste and hats back inside my panniers until they seemed to shiver, as if, were a clip to break, there would be a detonation, and a flying shrapnel of spare spokes, maps, pots and pans. I thought about all the kit that had proved worthless already – there was the electronic weather meter, which told me, with the press of a button, the temperature, humidity and, thank God, the 'dew point'. I had a wind-powered generator to recharge its battery too, a pathetic-looking plastic propeller that I'd tested for the first time yesterday. To generate any power at all, I'd need to be in a category five cyclone, the kind of life or death struggle in which dew points rarely spring to mind.

I made coffee, and, perched on a corner of the bed, cloaked in a duvet, switched on the BBC news. There was an item on the weather. The banner at the bottom of the screen declared this weather event 'The Big Freeze'. Scotland was entirely white and even parts of the south were buried under forty centimetres of snow. The temperature in Manchester had plunged to minus 17 degrees Celsius the night before, and scenes flashed before me on the TV, of highways consumed and closed, jack-knifed lorries and toppled vans. The army were assisting a pair of stranded motorists; they sat trembling beside a snow-covered road, looking spooked.

I reloaded my bike and pushed it outside, bisecting a pathway an inch deep in slush. A snowman was shaping up next door, fat and happy-faced, and out in the snowscape there were children at play. The weather had closed around eight thousand schools across the country and snowballing your mates to smithereens was apparently how to mark such good fortune.

On the road, I was trying to stick to where cars had broken the crust of ice when a few kids hesitated in their snowball fight. I looked up. An alertness had swept through the pack, as if lions had sensed a flicker in the grass of the Serengeti. I pushed down on my pedals and sped up, but it was far too late. The first missile smashed me in the ear. I took another to the neck, and felt icy water slip beneath my thermals.

The attacks were sporadic but they continued for several hours as I made my way south-east. The kids were *organised*, firing at will as I rode through Kent, flanking bridges and opening assaults from overhead walkways. Some were armed to the milk teeth with great caches of snowballs, others made soldier-style hand signals, and once, seemingly from nowhere, I heard 'Enemy 3 o'clock!'

Near Dartford, things got worse (they often do). A troop of kids broke cover from behind the Perspex of a bus shelter and bunched up like bowling pins. The lead pin was a red-faced sadist, four foot two of fizzing violence with the kind of sneer that suggested a prosperous future in corporate acquisitions.

'Get him in the face!'

'*What? No!* Wait! *Can't we all just …*'

'In the face! In the face! Get him in *his stupid face*!'

*

The Downs, my first taste of hills. I laboured upwards, cursing kilograms, my sorry legs, electronic fucking weather meters. But then … this wasn't entirely bad planning. I'd declined to train not simply because I was busy and wished to save time and effort, but for Science. Training would have upset the results of an important experiment.

A few months before, I'd pitched the idea to colleagues at St Thomas' of a study into the bodily effects of six years cycling. I was fascinated by how I might transform over the course of the journey, physically and physiologically, from an unseasoned 'before' to a granite-bummed 'after'. Dr Hart offered to preside over the experiments with what was, looking back, portentous enthusiasm. I took a seat in his office to discuss the plan.

'So I hear you're cycling around the world! Fantastic. We should get some background data.'

'Great! Sign me up for anything.'

There was a moment of silence. And then, tentatively, 'Anything?'

'Sure.'

Dr Hart looked down at his desk and scribbled something on a pad of paper. In retrospect, I believe he doodled a guinea pig, with a scalpel being driven through its brain.

First, I was sent to the respiratory research lab for 'Bilateral Phrenic Nerve Stimulation', which I understood was a way of testing the strength of my diaphragm using magnets, though I'd not bothered to look up the finer details before showing up.

There was something wolfish in the way the researchers watched me walk into the lab, and as I said hello, they seemed to be focusing less on my face than on my sternum and respiratory muscles. Clearly, there had been a dearth of such fresh and willing experimental material.

As I took a seat, one said 'Thanks for coming! It's great! Really great!'

'Right, okay,' I said.

'Now, try not to worry. No one's vomited before.'

'Vomited? But why would anyone ...'

'Nasogastric tube. We need to get data from near your diaphragm, and your oesophagus is the best way to get close. We'll get the tube up your nose in a sec.'

He waved a couple of bulbous things around. 'Magnets! For your neck.'

The other researcher chipped in now. 'Your arms and legs will spasm a bit when I hit this button. It's just a magnetic pulse, okay? But I should warn you, looks a bit like you're being electrocuted.'

'Looks like ...?'

He nodded, still smiling.

'Does it feel like it too?'

There was a horrid silence. Someone coughed.

I focused on not breaking his no-vomiting record as a tube was jabbed up my nose and left, as promised, dangling in my oesophagus. I was covered in electrodes and linked up to a machine with a hundred knobs on. The researcher stood over my shoulder holding a magnet in each hand, like a sinister puppeteer.

'Okay, ready? Here we go ...'

There was a violent pulse. My body snapped in all directions; my limbs were flung wildly but without any cerebral permission. Something in my head fizzled, possibly my cerebellum. The pain was visceral. 'Excellent,' he said – I must have spasmed correctly. 'Let's keep going.'

It wasn't over: during the next few weeks I found myself inside MRI scanners and an egg-shaped machine that analysed my body fat and composition. A muscle scientist, as eager as the rest, plunged a huge needle into my quadriceps. More magnetic pulses were shot into my legs until I winced when I walked and blanked out visions of cycling altogether.

*

But I was cycling, roughly speaking anyway. There had been quite a bit of pushing so far, especially when the snow piled up higher than my cranks.* Towards Chatham, the air reeked of burning rubber as I drew past vans and lorries stalled on icy slopes, a few of them abandoned. The light was steely now and the wind still full of snow, which nipped at my eyes. I'd planned to camp for free by the road, but this felt beyond me, and in Sittingbourne, the only alternative was to blast a fortnight's budget for one night in the Premier Inn. I was dithering in the road when a lady shovelling snow from her driveway called out brightly, 'What are you up to then?'

Travel is a famously good way of dispatching with lazy cultural clichés, provided, of course, that you let it. Perhaps I could start here, with the notion that speaking to a stranger in Britain is peculiar, if not transgressive. In southern, built-up parts of the country especially, it's something I'd been conditioned to believe only drunk and lonely people do.

'I'm cycling around the world,' I said, wondering if I sounded convinced.

'Really?! Well, you've chosen a …'

'I know.'

She was halfway to a smile now, impressed, perhaps, with the scale of my goof.

'Look, it's getting late. Maybe you could stay with us? Won't cost you anything! Come on, start again tomorrow. My

* Or, as I knew them at the time: 'long turny things with pedals on'.

kids are all grown up and we've got a spare room. Probably some cake around. I'll ask Roger. Roger!'

Travellers are forever gushing about the charity of the world, and I'd expected to as well, in Belize or Iran or Zambia, but not yet, not here. This was a glitch in the matrix, out of keeping, somehow, with British reserve, and with those sturdy, national traditions of gloominess and social constipation.

But there I was the next morning, in toasty, tumble-dried clothes, waving goodbye to Tommy and her husband Roger as they stood beaming at me in their driveway. They'd turned out not to be drunk or lonely or serial killers or Canadian, and in the absence of any other obvious explanation for breaking all the rules, there was really only one possibility left. They were just decent people, happy to help. This had been reaffirmed when Roger, a jovial, bearded man, rifled through a great stash of weathered Ordnance Survey maps that he kept beneath his armchair, muttering 'My wife wonders why I keep these!' For half an hour he'd advised me, in far more detail than was necessary, on the most suitable back roads to Dover. Bearded Englishmen of a certain age have a knowledge of B roads at least equal to that of Nasa satellites and if you search the homes of such men, you will always find enormous caches of these maps, stashed behind chairs and in the backs of cupboards, like pornographic magazines.

My tyres had a bit more traction now and the snow was well carved up, but it was still thick and smooth over the Downs. At last I reached the ferry terminal in Dover, where a woman was operating the vehicle barrier, or trying to, jabbing at a concealed button to get it to lift.

'I'm sorry, my dear,' she said, jab, jab, jab. 'This thing's a bit temperamental.'

'Bit like my wife.'

I turned. A truck driver was leaning out of his cab, grinning;

a great bloated grin, full of yellow teeth and jowls. The lady tried a withering look but she couldn't hold back a smile, and neither could I, so we all grinned away together, as if casual misogyny was just another thing to miss about Britain.

Inside the ferry, I locked up my bike and we heaved away from port. The white cliffs were soon dipping into the sea, and while home was still just minutes behind me it was also, I supposed, six years ahead. This seemed a vaguely discouraging thought, so I made for the other side of the ferry where there was a window with a view over the bow. We were churning through messy waves, the Channel seasoned with foam and whitecaps. As I searched the sea, I began to sense something else, something more than the dizziness of take-off, a flashback to those wide-open days when two giddy kids plunged through the winds of Patagonia. Call it the swirling, warm trace of adventure.

It had begun.

2

Horripilation

'Horripilation – *The bristling of the body hair, as from fear or cold; goose bumps.*'*
The American Heritage Medical Dictionary

It was hard to sleep in the cold, tent-flapping dark. Awake again, trembling, I put on more clothes and closed my eyes, expecting to repeat the cycle. Two hats, one fleece and a down jacket later, it was a tight squeeze inside my sleeping bag, my *four-season* sleeping bag, designed, presumably, by an optimistic Fijian who'd never seen winter in the French Alps.

Alpine cold is damp and infiltrating, and that was the worst thing. It soaks into your bones, hurts your chest, turns your brain to a slush puppy that can't plan or decide anything fast, like how to dress: yesterday's clothes were stacked beside me, frozen overnight into stiff, unwearable shapes. My thermometer had recorded minus 20 degrees Celsius in the early hours and the roof of my tent bowed inwards now as snow continued to pile on top. I turned over in my sleeping bag and found

* Horripilation is derived from the Latin verb *horrēre*, meaning to stand up, bristle, shiver or shudder, hence the etymology of a *hairy* situation. 'Horror' was used in a medical sense into the twentieth century, meaning a shaking, shivering or shuddering, as in the cold fit that precedes a fever, a chill of less severity than a rigor.

my water bottle grinning back. Liquid had expanded to ice, splitting the metal and opening a mouth that was drooling melt-water. I knew why the world was mocking me. There would be much higher mountains ahead, more extreme cold, yet here I was, toughing it out in dinky, safe Europe, the most familiar, and least formidable, continent of the lot. Ha ha fucking ha.

Gradually, I took down my tent. My gloves were still too rigid with ice to wear so my fingers burned with cold as they stuck to the metal poles. I tried to yank the poles apart at the links, but nothing worked. Not bending, not twisting, not flinging them to the ground in a childish rage and screaming, 'Come on you fucking pole! Fuck! Pole fuck! You fucky pole! Pole!' Eventually, I fired up my stove to melt the ice and free them.

After the ferry had docked in Calais, I'd cut a long diagonal across France, rounding Paris, plunging south via Champagne countryside and through Dijon. France had been just one big dirty-white fridge so far. The fields of Brittany were defined only by hedgerows, and near Étaples I had to squint to find the pale headstones of the war dead. I was in the south-east now, close to Briançon, France's loftiest city, hoping to climb Col de Montgenèvre, which scaled the border into Italy and promised even more horripilation.

That morning, I stopped at a café and sat hunched over a cup of coffee, unable to recall the precise logic that was driving me upwards, into even colder air. So I ignored Maude, the waitress, when she asked me why I was cycling during the winter. It was happening, it didn't help to ask why. Perhaps it began with some yearning to prove myself, but I'd misjudged the brute force of the winter, or myself, and right now, with a snot-icicle under each nostril, my over-ambition felt blatant. I could have been riding smooth bike paths beside sleepy rivers but I'd vetoed this kind of common sense because it sounded like a half-term holiday, too gentle a foundation for the adventure I had in mind.

I'd like to say that there was a pep talk – there wasn't. I wish I could say that I tapped into some inner wellspring of strength and tenacity, like those solid adventurers in my books, that I turned my eyes to distant peaks, growled at them perhaps, and persevered. And I wish that, now, I didn't have to write that I left the café, pulled a U-turn and set off for Nice.

Warmth flooded back to my hands as the road declined towards the coast. There was a moment of intense joy when I patted my face to find not frost, as expected, but actual eyebrows. And despite the nagging sense that I'd backed down, cheated even, life felt much easier on the Riviera, supping lattes and scoffing warm croissants. The deluge continued however, albeit in a different state – snow had turned to rain. It was drilling into me, or drifting down in a fine spray, misting the road, but it was ever-present and ever-cold as I followed the coast into Italy.

I cut inland, crossing the top of the boot, shooting vaguely for Venice. In the afternoon, a figure appeared in the deluge, tight against the edge of the road. His face was hidden in the hood of a raincoat and a backpack tipped him forward slightly as he walked. He turned when I stopped alongside him and a youngish face, lightly bearded, burst into a smile.

'Hey, hey!' he stuttered. 'A travelling man!'

'Where are you heading?!'

He threw an arm over my shoulder, 'Mongolia! Allons-y!'.

Unsure if this was a joke or a lie or wishful thinking, but intrigued enough to find out, I got off my bike and plodded with him through the rain.

Matéo was a sculptor tramping south from his home in Brittany, stopping every now and then to build elaborate cairns from stones he collected along the way. He wore a piece of string around his neck threaded with the red beak of a small bird. A baguette protruded from his pack like a sword. Why

Mongolia? He shrugged as if to say that an adventure had to end somewhere – and Mongolia is, after all, a byword for distant and extreme.

As we passed a café, I offered to shout Matéo some tea. Inside, he ripped off his socks and boots, slung them to the floor, rested back in his seat, and slapped his bare feet on the table. They steamed horridly. A waiter appeared, shouting at him to get his feet down. Shooting daggers at the waiter, Matéo did as he was told and then leaned in and whispered, conspiratorially, 'Careful. Zat guy is craaaazy!'

Matéo carried no phone and no map, he simply ambled towards the sunrise, advancing slowly, and vaguely, on the wide Mongolian sky. He was consumed by this dream now, and in its pursuit, he rarely paid for a bed or even food. Behind a nearby supermarket he showed me how, diving into the big metal bin, head first. Seconds later, apples and cheeses jumped out like pieces of microwaved popcorn and I ran about, arms akimbo, catching what I could. As we sat together, stuffing our faces, I ached for his nonchalance. Matéo was undeterred by the rain or cold, by his miserly means, by the slow-motion chaos of his life. He was inspired by uncertainty, like I had been once too, but its magic was lost on me now, and anxiety pecked at my days. Perhaps I should have known better. Working as a doctor is a lesson in how double-edged uncertainty can be; it hides beneath so many of the adventures and tensions of the job.

Alone again, fuelled by focaccia, I made good progress beside the broad Po River, the land impeccably flat for miles. As fields went by with the monotony of paving stones, I killed the hours, unmoored, slipped off into a stream of thought. In my withdrawal from the galloping life of London, the cut and thrust of hospital medicine, I'd gone cold turkey. No computer, no phone, no pestering notifications. I read more than I ever had – a book a week – alone, with nothing else to fill the void between dinner and sleep.

Already, cycling had become a sort of cover for dwelling on questions and on life, but I'd not experienced such depth of time before and the solitude felt challenging at first. I unpacked the years behind me on days like these, letting memories resurface, the good and the bad. At times my mind felt like the end of a Super 8 projector reel, looping and flapping without end. But each passing day felt easier than the last, and there was a feeling of release sometimes too, especially when I thought about the decisions I had to make – there were few, most carried little weight, and they only impacted me. Previously, Wednesday mornings meant deciding on diagnoses and treatments, and sending patients home 99 per cent or 95 per cent sure that it was the right thing to do. This is the millstone you learn to live with as a doctor, the tormenting sense of accountability, the late-night palpitations as you wonder whether forgetting one of the seventeen causes of epigastric pain will have consequences you will never forget. These thoughts seem melodramatic as you progress up the ladder and get better at your job, or at least more accepting of its burden and your own limits, but they cloud the start of every career.

I recalled my patients, which helped pass the time when I was bored of podcasts or of fields. Some memories – the hardest and most frightening ones – arrived as still shots. Like the teenage girl in the throes of crystal meth withdrawal. She'd broken her foot after kicking the walls of her side room. Now she lies, still at last, forcibly shot with Valium in a room splattered with her own blood and puke.

'May I?' asks a man. His limbs are massive and bruised, like bloated aubergines; they ooze fluid. His heart is failing. I nod and he turns on a small portable speaker by the bed. 'Bach', he says, 'helps with the pain', and his free hand is swishing through the air as I unsheathe a needle to siphon off his blood. Not a flinch as I toil to find a vein.

My successes and failures as a doctor came to mind too – small moments I am proud of, and the stinging, human catalogue of could-do-betters that every doctor owns. It's occasionally a comfort to think that I might have contributed to saving a life, and when a patient dies despite your hopes and diligence, well, it's important to remember that healthcare is very much a team effort.

*

With my mind on its own expedition, days ticked by and I got to the Adriatic coast sooner than I'd expected. It helped that the days were longer now too, and blue-lit, the sea and the sky in pleasant symmetry until dusk. It might have been a happy time, if it wasn't for my knee. Over the last few weeks, the left one had grown swollen and sore. I could still cycle but it clicked, there was an unstable feeling when I walked, and I could feel a small lump, ominously mobile, within the joint. *Huh.*

This was particularly frustrating given that, physically at least, I'd begun to adjust well to the long days on my bike. I'd lost a tenth of my city-boy blubber despite an unbridled appetite. Eating – ecstatic, tremendous, frenzied eating – is an essential ritual of cycle touring and I was now helplessly in thrall to food. One of my front panniers was now mostly biscuits (I'd made space for them by throwing away life-saving components of my medical kit). Sometimes, eating simply happened, and I was only semi-aware that it was taking place. I'd find myself slumped beside supermarkets, burping, a taste of salami on my lips, flakes of *byrek* over my chest, thinking *Shit. It happened again.*

Mostly, I had just glum horses and the odd tired Mercedes

for company in Albania. The roads ranged over forested hills, sparking an occasional, lancing pain in my knee that I decided to deny was happening. As dusk fell near the town of Elbasan, I was just setting up my tent on a village football pitch when a man introducing himself as Zef, looking disappointed by my lifestyle, invited me to stay in the warmth of his home.

Save for the odd crucifix, it was a plain, white-walled place. His wife, sister and three kids sat around me as I ate my fill of sausages, eggs and gherkins. Zef removed my shoes, slipped a pair of slippers onto my feet and offered me a cigar for dessert. We talked, in a way. I understood only what could be most easily mimed: that his ten-year-old son, Albert, wanted to be a boxer when he grew up (or an Olympic swimmer, if that round-house punch was front crawl). The dog was a pain in the arse and ate too much, that was clear. Zef's sister understood a little English, but she was profoundly deaf so she scribbled down questions on notes that were slid towards me, and a circle of faces turned to me hopefully as I considered each one in turn.

After my job and country had been revealed, a note asked 'Are you shameful?', and given that it had been written with the help of a dictionary, I decided that she'd meant shy. Under the circumstances, it was hard not to be. I said maybe, a bit. Another arrived. 'Are you happy?' I thought about it: my knee ached, true, but the worst of the cold was behind me and I hadn't died an adventurer's death, in a snow cave, scribbling my last words to loved ones. My new life was simple and self-contained, days stripped to the bone and time galore to turn things over in my mind. Something about the action of cycling helped me to think – the whirr of spokes, the flow of parting air. I wrote 'yes!' and felt adamant. Another note arrived; it read: 'Princess Diana. Accident or murder?'

*

I crossed from Macedonia into northern Greece, and it was time to accept that my knee was unlikely to be healed by Tubigrip and hummus. It was much worse now, warm and tender and swollen. So in Thessalonica, I shelled out my month's budget for an MRI scan, which was horribly conclusive. There, on the computer screen, was the dark inverted dome of my femur, its contours smoothly anatomical until a sudden bite mark, 11 mm across. A chunk of cartilage had opted to go it alone explaining the rogue lump inside my knee, known colloquially among orthopaedic doctors as a 'joint mouse'.

Simply put – and you'll have to excuse me, this might not sound very medical – a bit of my knee had died and fallen off. The medical world gives a collective shrug when pushed on precisely why people get *osteochondritis dissecans*, the unfortunate problem of cartilage coming loose spontaneously, but in my case, it was probably related to repetitive microtrauma, which is what can happen when an untrained, flabby man turns very suddenly to trans-global cycling. The radiologist didn't need to say it, but he did anyway.

'Go home. Only a surgeon can fix this.'

Certainty, at last. But I was still unsure if my knee would recover enough to take 50,000 miles on a bicycle afterwards. Perhaps that line around the world was just that, a line, a hopeful squiggle, never to be consummated. I'd spent years summoning the courage to jump, wrapping up one life to unwrap the next, and maybe now all those daydreams were for nothing. 'I went to the woods because I wished to live deliberately', wrote Thoreau, and I'd felt equally convinced by the wilfulness that imbued my own adventure. I would be gutted to let go, and drift back to night shifts and British skies and beans on toast.

For now, however, heading home was a reality I'd have to live with, and though it was small consolation, at least I could exploit professional connections. After emailing an orthopaedic surgeon at St Thomas' and sending the scans, an operation was scheduled. I resolved to limp over to Istanbul before returning home. Then, following surgery and some weeks of physiotherapy, I'd fly back to Istanbul to continue biking, provided my knee could handle it. It was a simple plan, complicated by an Icelandic volcano with a name to match the trouble it was causing. When Eyjafjallajökull erupted and cleared European skies of planes, I was told to wait a few weeks or to find a plan B.

*

Stacked under a bed in my mum's house in Oxford lie strips of cardboard. They are old and frayed memorials to the hitch-hikes of my teenage years, to a nascent wanderlust, to plan Bs. Nowadays, I like to think of them as warning signs of a bike ride around the world. There are marker-penned petitions for 'M1 north' and 'Milton Keynes', even a playful 'Anywhere', which I held up for an hour on a quiet stretch of road near Bicester, and which, from my savvier vantage point today, reads like an exciting opportunity for a passing sociopath with a taste for kids.

Still, I hitch-hiked often as a teenager, with only a few mishaps. Sometimes I had no destination in mind, I was simply content to move; to sample those small, fleet surprises that jumble any journey. Hitch-hiking endorsed human beings, and it was a gateway drug that left me busting for somewhere more exotic than Basingstoke. It could be an exercise in profiling, too, and I quickly discovered which tribes were most likely to offer me a ride. You want the socially adrift and lonely, eccentric sorts, free spirits, beards. Back then, I looked for mums and dads as well – they were more sensitive to my age. A surprising number of born-again Christians plied British roads when I was growing up, all convinced that teenage hitch-hikers were ripe for religious conversion. Do I want to be bathed in the light of our Lord? *Yeah, sure. And then ... could you drop us by the Little Chef in Doncaster?*

One day, Brian, a fridge repairman en route to Newbury, steered his Saab onto the verge and flicked open the passenger seat. After twenty minutes down the A34, Brian's phone rang and when he answered it on speaker, a slow, liquid voice entered the car.

'Hey, big man.'

Brian shifted in his seat.

'Guess what I'm doing with my ...'

'Oh Julia, hey! Got a hitch-hiker, babe!'

Brian gave me a nudge to speak up.

'Um, hi Julia,' I managed.

There was a muddled silence followed by the hum of a dead phone line and a titter from Brian. I was still grinning when we reached Newbury. From religious nutjobs to the unfathomable Julia, it was a powerful lesson: you never knew what travel might throw up.

*

Humping a rucksack, having stashed my bike with a Turkish friend, I set off from Istanbul to rediscover my love of hitch-hiking. It took eight lifts to breach the limits of the megacity, the last courtesy of Apo, a Kurdish perfume seller, the cloying scent of 'Candy' filling up his hatchback. An hour later, outside Tekirdağ, I was sitting in the cab of a truck full of car parts, next to Hussain, a blue-eyed Turk who chain-smoked Winstons. Aside from Turkish, Hussain spoke only broken Italian, and since I spoke broken Spanish, for two whole days, this was a messy bargain that would have to do. A six-foot-six basketball player delivering coffee machines to Switzerland dropped me at Le Havre and from there I travelled to Portsmouth by boat. From outside the port, another trucker, Greg, was my twenty-third lift from Istanbul. He dropped me in Oxford, twenty yards from my mum's front door. Unfortunately, she wasn't behind it, so I plodded down the road and spent a nervous night camped in the park where I'd been head-butted by a thickset thirteen-year-old girl as a kid. I'd slept better in the forests of Albania.

Soon afterwards, I was in Guy's Hospital staring up at an anaesthetist, my left knee marked with pen so that they didn't open up the wrong one. When I awoke, foggy with morphine, I'd been wheeled into the bay of an orthopaedic ward and parked between a double amputee and a man with a spinal fracture who had to be log-rolled every time he shat himself, which of course gave my troubles all the magnitude of a paper cut.

Twelve weeks after the surgery, and with plenty of physiotherapy, my knee was hinging usefully. Istanbul was soon sprawled before me again and, though excited to be back on track, I knew that I'd have to start carefully, spinning in low gear. Going slow proved to be the only option anyway; it was August, and there was a close, molten heat that checked my pace more than any twinges did. Inland, exposed on the Anatolian plateau, Turkey was burnt to rust and butterscotch. Everything appeared camouflaged now: the hoopoes fluttering between trees, the parasitic dodder vine invading the fields, the hulking Kangal shepherd dogs, too wasted to give chase. With my lips of burnt rubber, I envied the kids sploshing about in the stilted irrigation channels, and found fleeting relief when trucks rushed past, too close, compromising my life in lovely gusts of hot air.

I'd gained altitude slowly through central Turkey and so it was all downhill to Syria, a dash through the Taurus Mountains, waves of pine forest beside me, the land flattening at last into groves of olive and pistachio trees near the border. This was 2010 and the Syrian tragedy had yet to play out. Sixteen months after I pedalled into Aleppo, the city would shake under aerial bombardments, house to house fighting and slaughter. At the time, I had no sense of such a precipice and the triumvirate of Hafez, Maher and Bashar al-Assad, dressed in shades and military regalia, stared from photos spread across the rear windows of cars in calm assertion of the status quo.

Because bicycles are slow – at least the way I ride them – big countries, despite their subtle diversities, grow familiar and so it can be jolting to leave them behind. Turkey is big, three times bigger than Britain, and bigger than you imagine until you mark your progress over a map. In my slow acclimatisation, Turkey established Syria as somewhere very new indeed.

I strolled through the souk in Aleppo, a vast covered maze that smelt of olive oil and soap, where young men with slicked-back hair sipped fruit cocktails, puffed cigarettes and played it cool. I'd learnt a hundred or so words of Arabic but I was unprepared for a rich dialect of hand signs used in place of words. A breakthrough: chin back and click of the tongue means 'No'. Later, I would learn that there are countless interpretations for holding a hand outward with the palm down and then suddenly twisting it up, among them 'what?', 'why?' or 'how?', yet despite this general ambiguity, it always seemed to ask specific questions of me. *What are you doing in Syria on a bicycle? Where are you going?*

I was hyper-vigilant here, wide-eyed, presumptuous; believing, *knowing*, that every sight, smell and sound was typical of Syria. A delusion, of course, and sadly one that works in reverse – there are probably now several men in Aleppo who believe that all British men have shit beards and mayonnaise stains on their clothes. Culture shock isn't always uplifting; it can be grim and overwhelming too. It's even been taken for a psychopathology from time to time (but then, what hasn't?). Paris Syndrome is perhaps the quirkiest example of this, a transient mental disorder* said to occur in around twenty Japanese visitors to Paris each year. It has been traced back to the wives of diplomats in the 1970s, who arrived in the city with a particularly clichéd Paris in mind. Sufferers of Paris Syndrome are said to grow paranoid and delusional. They can hallucinate and panic, reel and even vomit, all out of sheer distress: Paris is not what they expected: not twee perfume-scented models, joie de vivre and lovers. There's homelessness, sulky waiters and Kentucky Fried Poulet.

*

As I headed south through Syria, flat desert on both sides of the highway, only the occasional, abandoned 'dead city' reformed the view. Aptly perhaps, it was my thirtieth birthday on the road to Damascus, and although I was on my own with no celebrations in sight, my knee felt strong and I was going swiftly on my way, so I felt none of the angst that a new decade can bring.

I was flagged down that afternoon on the outskirts of a village, somewhere between Aleppo and Homs. Tariq was young and smartly dressed, and he spoke good English as we sat out of the sun, tearing chicken apart with an older, squat man called Mustafa, something like a village headman, who

* Though not, I should say, an official one as defined by *The Diagnostic and Statistical Manual of Mental Disorders*, which arguably includes weirder and shakier 'syndromes' than this.

had four wives and eighteen children. Mustafa dispatched with the chicken fast and mercilessly, leaving a slimy glaze on his moustache. He muttered something and lent me a big, lopsided smile. Tariq translated. 'Mustafa says he loves you. He wants you to have three kilograms of meat.'

I pointed to my bike. 'No room, mate. But look, tell Mustafa thank you. It's actually my birthday today and it's been—'

'Birthday!' Tariq leapt up, his eyes bulging, excited. There was no way to stem the flood of hospitality after that. Aftershave and hair gel were pushed into my hands; I was hustled off for a shower. Tea, more tea, never enough tea. A sequence of women gathered up my clothes and disappeared to wash them. Mustafa invited villagers over, scores arrived as the day drifted on, and I was paraded in front of them, dressed now in a white kaftan robe and a red and white scarf, a *shemagh*. When evening fell, Tariq's brothers dashed off to make a sheet tent for me beside the house. I dozed off that night warm and grateful, and oblivious to the weight this memory would bear when Syria imploded in six months' time, or later, when I met young men among Syrian refugees and imagined Tariq homesick and fugitive too.

3
Peripheries

I first met Nyomi six months before I left home and for a short while we'd shared a pokey flat in a charmless part of London. What struck me – what struck most people about Nyomi – was her refusal to conform. Ny distrusted rules and tried not to let anyone's expectations steer her away from an instinctive, headlong approach to life. Enduringly skint, she'd open cans of cider under pub tables and cadge stale sandwiches from Pret a Manger for lunch and dinner (which, of course, made a mockery of the shop's name). She had a habit of rescuing broken bikes discarded around London, retrieving them with love, like they were orphaned kittens. She was resourceful and radically frugal and powerfully different to anyone else I knew.

At the time, I was planning the journey ahead, and Nyomi took notice. One evening she zoomed into my bedroom with her bright-idea face – it was an expression that caught her at least once a day. Drawing herself up, she readily announced that she would like to cycle the length of Africa with me – and for Ny, practicalities were always boring, minor things. What makes the best kind of travel buddy? If it's an undaunted, happy-go-lucky streak, as I'd always assumed, then Ny was probably perfect. It was set.

After Syria, it took me a few weeks to cross the deserts of Jordan and Sinai, though navigating the bursting streets of Cairo, the warbling muezzin clashing with car horns and barking traders, felt like an equal task at times. I found Ny at the airport and we were soon unpacking her bike box, and pulling out her single-skinned dome tent, begged from a friend. The label called it 'A budget-priced tent for sheltered summer use, aimed at youngsters seeking their first camping adventure, perhaps in the back garden'. So that's how we began our journey together: shopping for a better tent (a marathon task in Cairo, a city of twenty million people, of which seven or eight are apparently campers) and stuffing ourselves daily with *koshary*, a mix of rice, macaroni, lentils, spiced tomato sauce, garlic vinegar, chickpeas and crispy fried onions, a food apparently invented for skint, carbohydrate-junkies like us.

We left Cairo with half-baked plans. We were going to stick to the eastern side of Africa at first, switching to the west, further south. The river guided us away from the overloaded city, and soon the roadside was lined with prickly pear. In the gaps I glimpsed the Nile, coppery and calm, and the odd felucca, slipping in and out of a thin surface mist. Palm fronds defined the far river bank, ink-black against the blush of sunrise. Ny was cycling ahead in a large brimmed hat, dreadlocks dangling from it, tied with yellow and green twine. Spring onions and cucumbers were stuffed under her bungee cords, a catapult was strapped to her handlebars, and a sign cable-tied to her rear rack said 'I DON'T BRAKE FOR ANYONE'. She was a contradiction in motion: a dangerous, grinning, misanthropic hippy, all freckles and pent-up road rage.

Perhaps this was why, at a roadblock, a policeman gave us a wary once over, and then insisted that we were escorted by a police car. When we questioned the necessity of this, he pointed to a field of sugar cane hissing in the wind and, holding a pretend rifle to his eye, warned us of the sniper within. No amount of eye-rolling helped our cause – this was protocol – but once we got over the presidential touch, it grew irritating to ride before a grunting car until we were passed to another police officer to repeat the job in a new jurisdiction, like batons in a nationwide relay. The loitering police presence meant that we couldn't steal away and rough camp at dusk. Confined to a hotel each evening instead, we rose two hours before dawn, packed up quietly and set off alone, hoping to lose our tail. But they would catch up to us eventually – you can't hide a touring cyclist in Egypt. When we weren't trailed by the police, we were trailed by a frenzied wave of Egyptian boys, who mobbed us when we stopped to repair a puncture and lost their shit entirely if there was a chance of getting photographed. Lost some cyclists? Just follow the screams.

It was a dozier world the further we got from Cairo and eventually the police let the loping camels be our chaperones instead. We were closing in on Sudan now, which was spread beyond Lake Nasser, a vast reservoir created by the Aswan High Dam in the 1960s. The boat left when it was fully loaded, with people and pots and rugs and washing machines, well past the advertised departure time. When punctuality can pass for impatience, you recalibrate your watch. There were just a dozen foreign travellers in deck-class, a small, sunburnt island – all vests and shorts and dovetailed airbeds – with everyone else mummified in hats and coats. The Sudanese asked us quietly about our travels, one at a time, while the Egyptians went for a fun, racy, mass interrogation.

After the boat docked, we pedalled a short distance to Wadi Halfa, a dusty, low-slung town with that end-of-the-road quality, shops stocked incongruously, bananas sharing shelves with teddy bears and tyres. We stocked up and moved on, cutting through desert. The road was smoothly paved and shipped little traffic; we were spurred on by the wind rushing south. Around midday, as the sun grew severe, we ducked into a tubular, corrugated metal culvert beneath the road, and whiled away two hours, snoozing and guzzling murky water collected from clay pots by the road. We were back on the road as the afternoon cooled, and by evening, the Sahara was a sharper, prettier place. The sand dunes were rust-toned and rippled now, and the sunset had left a smudge of wine to the west. Soon, Nyomi was just a silhouette up ahead, sliding along beneath a pinpricked sky. It would have been a faultless picture of Saharan serenity if she hadn't been rapping along to a tune in her headphones, but it felt nicely absurd, out here, to hear The Notorious B.I.G:

'On the Lexus, LX, four and a half
Bulletproof glass tints if I want some ass
Gon' blast squeeze first ask questions last'

*

We shifted tack after Khartoum, moving east towards dim mountains. When we crossed the border into Ethiopia at Gallabat, time shuddered – we no longer inhabited 2010, but 2003, for we'd passed into the Ethiopian calendar where the month was now *Taḫśaś* and the official East Africa Time Zone was largely ignored: for most Ethiopians, the day begins at dawn, not midnight, and the time is the number of hours that have passed since sunrise. Beyond the temporal, there was a new, vigorous feel to the land. From desert, blanched by sunlight for most of each day, we transferred to flowery scrubland, where rollers, with plumage as blue as a kingfisher's, flashed between trees. Small sounds had ruled the desert night: insects, scuttling beneath my tent, loud against the near silence, sounded like machines, but the dusk was raucous here, and richer. There were shepherds whistling, people calling in Amharic, strange birds twittering, and whips on the hides of oxen. Animals woke us well before the sun, with a warring dissonance of crows, barks and brays. There were murmurs, too. One morning, I opened my eyes to almost twenty children; they had unzipped my tent and, packed together, wonderstruck, had been watching me sleep.

It was partly the kids, and partly the thumping beauty of the Semien Mountains that made those first days in Ethiopia so demanding, and so gripping. Night or day, the children came like an avalanche, screaming and surrounding us four-deep. There was actual violence of course: the little scamps fired at us with slingshots, and a painfully keen aim, the stones clinking off our bike frames when not inflicting actual bodily harm. This was gameful rather than malicious, the Ethiopian equivalent of chucking conkers at a stranger's window, my own particular prank of choice, back when I was a boisterous ten-year-old on the block. We'd been expecting it too – getting pelted with stones in Ethiopia is written into cycle-touring lore. The hair of some kids was shaved except for a single tuft, a hairstyle known as *Quntcho* and mythologised as a sort of handle for angels to pull mischievous kids out of trouble. It was hard to resist snatching one myself.

As we tackled the Blue Nile Gorge, more children appeared, running – always running. With the easy bit done, 1,300 vertical metres had to be painfully regained, and soon a few kids were pressing their hands to our panniers as we climbed. The game caught on and Ny was barely bothering to pedal as a whole gang of shrieking ten-year-olds pushed her up the gorge (while filching what they could from her rear panniers – underwear, mostly). A better cheat, then, was to grab onto the back of one of the slow, rusty trucks that climb the hills only marginally faster than a cyclist can, though at times our extra weight slowed them to a standstill and we were forced to release our grip. It was just one more mad fact of our journey through Ethiopia: we could be propelled up mountains by children, and we could arrest fifteen tons of metal and grain.

*

My map suggested only one traverse of northern Kenya, and it wasn't pretty. The Moyale Road, while paved in parts, was mostly a fudge of sand and bike-snapping corrugations, crossing a poor, marginal region, reputed to be the roaming ground of bandits known as *shifta*. Armed guards accompanied overland trucks through here and already I'd met several people who'd been physically damaged on this road, including a man who assured me that he'd been shot in the shoulder through the windscreen of his truck. He pulled down his T-shirt, and, reluctantly, I examined the scar.

'You see? The *shifta* will take everything you have and then they'll kill you. They'll take your shoes.' He made this last statement with outrage in his eyes, as if stealing the trainers from a corpse was an atrocity too far.

I looked down at my sandals, a snug fit. I'd always been fond of their solid soles. I studied my map for other roads. One caught my eye. It wormed west through the tribal lands of the Omo valley in south-west Ethiopia and then formed a dead end by the Omo River and the border with Kenya. Further south, the blankness was perfect until you reached a long, blue lake marked Turkana. Joseph Conrad wrote about the call of blank spaces on maps, and I had to admit that there was an other-dimensional quality to the vacancy, as if the river marked the end of the world, or a portal to somewhere new. If I could cross the Omo River and find a trail, I'd break through a leg of the Ilemi Triangle, a mean and dust-riddled place, mostly, though perversely, disputed – Ethiopia, Kenya and Sudan all claimed the badlands as their own.

But fear can be inspiring, at least, in not too high a dose. I'd found a blog post from touring cyclists who'd travelled this route a few years ago, a flinty couple hardened by half a decade of remote bicycle travel. But even they'd sounded put off. They wrote of fierce heat, raging easterlies, vipers, bursts of tribal conflict and the near constant threat of getting lost. Such stories should have put paid to my daydreams, but in a way that concerned me a little, possibilities like these made me feel wonderful. I felt pulled towards Turkana now, like a petulant child told not to play in the woods.

In Konso, Ny threw her bicycle onto a bus bound for Nairobi, eager to meet her boyfriend who was flying in from the UK to visit. We arranged to rendezvous in a month, and with two kilos of rice and a litre of honey, I began nosing into the Omo valley, alone.

The bush was thicker now, and termite mounds, high as houses, rose in the distance. Twice, antelopes spooked. The sandy trail was scuffed by four-by-fours belonging to NGOs or the UN. Some of the villages suffered fleets of tourists stalking about with zoom lenses. I passed their subjects from time to time, Hamer women crouched by the road, dressed in ochre, leather, copper and cowries, hawking their tribalism: photos for money. Commoditising tradition had brought in cash, developed a little infrastructure and had perhaps reinforced the traditions themselves (including the ritual flagellation of women and girls), but often the result was a human-museum quality that everyone appreciated was strange. I was a tourist too of course, although I didn't exactly embrace the fact. Tourists were naff people: shrill, bovine, jowly weirdos with knapsacks and visors. I was An Adventurer, at least that's what I told my mum during our strictly scheduled phone calls.

Outside a few huts, I was surprised to see another touring bicycle, though with no sign of the rider. A broom handle was bungeed to the bicycle's rear rack, which I supposed filled in for a kickstand. There were three large water bottles strapped to the frame and scraps of electrical tape unravelling about the cables. Unkindly, I hoped that the bike's owner was as ruined as me. My T-shirt was a leopard print of sweat patches and grime. My Lycra shorts had so many holes that they'd become unsexy fishnets and my over-shorts had holes in the crotch. Two days ago, I'd put my entire leg through one of these rips when trying to find the leg hole. My tyres were smoothed of grip, my cooking pan was badly charred, and my towel ... Jesus. My towel looked like the type of thing you might wrap a dog in, if the dog was bleeding and needed to be taken urgently to a vet.

A man appeared from behind a hut, legs so wiry that his Lycra shorts hardly clung. Beneath a cap embossed with 'Che Guevara' was a grey face, rugged in stubble. His scowl looked glued on, but he smiled and wrinkled his eyes when I called out 'Hey!'

'Slow going, huh?' he said. I nodded. He scratched his belly and sweat rash.

'You heading this way then?' jerking his thumb at miles of dwarf scrub and sand. I nodded again.

'I came that way. It's hot as hell, I tell you. Tribes are fighting. They're killing each other. I saw body bags.'

He seemed content to let that linger for a second or two, and I wasn't sure how to respond. He walked over, kept coming, until his face stopped too close to mine.

'You don't mess about out there, okay?'

'Okay'

'Bring more water than you think you need or you'll get fucked up. And mark the sand with something so you can get back if you need to. Get lost out here my friend, you're dead.'

'Right,' I said, smiling at the Hansel-and-Gretel-ness of it. Mark the sand with what? Super Noodles?

He glared.

'It's a shame about the traffic,' I tried. Nothing.

'I'm Stephen, by the way.'

'Jörg.'

'Where have you been riding?'

'I cycled across Africa three times already.'

I looked again at his broom-handle kickstand, his scuffed-up panniers. I wondered about his towel.

'How long have you …'

'Years,' he cut in, bored. 'I'll die on my bike.'

I'd have laughed, but he made it sound like a fact.

On the face of it, there was nothing about this neighbourhood that tempted a touring cyclist, but we'd both chosen to ride here, so maybe we had more in common than sweat rash and a dismal dress sense. Maybe Jörg shared my own desire to feel remote and half-stranded. I was after an adventure, but what was that? The travel writer Tim Cahill joked that an adventure was never an adventure when it was happening, but only in hindsight, it was 'physical and emotional discomfort, recollected in tranquillity'. Or maybe 'adventure is just bad planning', as Roald Amundsen asserted. I had no reliable map, no GPS, and the trails ahead were as transient as the winds and the people; shifting gutters of sand, sketched anew each day by weather and livestock. By Amundsen's contention, I was an adventurer extraordinaire.

Jörg spoke of other, old adventures, of tracks he'd suffered in the Congo, Chad, Niger, about an infection that had almost cost him his life: 'Wounds don't heal well out here, they rot. Lucky I still got a leg left.'

I wondered if risk was key to his sense of adventure, to feel at risk and to risk it all for the sensation. 'What gives value to travel is fear. It breaks down a kind of inner structure we have', said Camus, and I was beginning to sense he was right. Jörg travelled alone, as I did now, and without Ny, I'd noticed how solitude modifies travel, inflating the risks, sharpening your senses, holding you, for small, important moments, in a state of sparkling fear. I left Jörg to his quest, whatever that was, wondering when resolve tips into obsession, or if he would know when it did.

I hoped that something had been lost in translation outside Omorate, the last village before the Omo River. A sign said 'Welcome. Value your life'. There was no bridge here, just a 20-metre span of dawdling, muddy water, but at the river bank a boy in a dugout offered me a lift across. I removed my panniers, and handed my bike over too; he held it upright and, at the same time, impelled the canoe to the far bank with thrusts of a pole. I watched, everything I owned drifting away, but perhaps he sensed my anxiety because he returned for me quickly, with a big grin.

By hopping the river, I hadn't arrived legally into Kenya – there was no border post here – and I hoped that a compassionate immigration official, if that wasn't an oxymoron, would grant me an entry stamp when I got to Nairobi. On the far bank, I took stock. There were few signs of life. The desert was long and lion-coloured, with a few tussocks here and there, and a remote row of acacias in silhouette.

Turkana is Kenya's largest county and its most desolate. It spilled with life once, but the centuries have not been kind, desiccating, wind-scraping and baking the land. It is a thorny world, a struggling world, of ancient granite and recent lava flows. Even the most dogged of life is deterred. Often, there is little here to guide you, and it was only when I arrived at the goat's carcass for the third time that I submitted to being lost. Around me, vague dunes were melting into the sky, the wind whipping up spindrift, and soon the afternoon had the shapeless threat of twilight. Scudded by sand, my wheels sunk to the rims, as if I was drowning in desert – though it felt more intentional at times, as if the desert was dragging me down, and drowning *me*. A little fear crept in and spread, and when the trail faded to almost nothing, I felt overwhelmed, but there was something else here too, something inspirational, and a painful sense that I shared Jörg's fascination with risk.

I'd hauled water from Ethiopia, lots of it. The water in Lake Turkana, to the east, was too salty to drink, and the *luggas*, the seasonal water courses, too unreliable. Twenty litres was the most I'd ever had to carry on my bike; it sloshed in tethered Coke bottles and a ten-litre water bag, and the combined weight of my bike and load now outstripped my own. Turkana's aridity is famously preservative, and maybe in 1.5 million years a future archaeologist would scoop me up, string my skeleton back together and hoist me up for a museum display, like the Turkana Boy, the most complete skeleton of Homo erectus yet discovered. No doubt my towel would survive too, retaining its horror.

In the late afternoon, a letter 'i' in the distance resolved into a Turkana warrior, an assault rifle hanging off his shoulder. His lower incisors had been bashed out; a Nilotic tradition thousands of years old. There's a theory that the practice originated because it enabled those in the grip of tetanus to be fed through the gap in their teeth. I wasn't convinced. It seemed too easy to call every tradition adaptive, and perhaps this was just another human fad, as practical as a mullet or hipster beard. The man pointed at a two-litre water bottle in my holder, half full. I handed it over and he tipped it upright, drained it, then wandered away, his silhouette lost again in the desert's dream-sequence haze.

Turkana was at the limit of every kind of landscape I could think of: national, economic, climatic, political, habitable … Under colonial rule it was a closed area where expatriates needed permission to enter, the British judging it treacherously remote and hostile, and Turkana warriors were reputed to spear strangers who ventured within. Over the years, there had been various arrogant and failed attempts to 'pacify the natives' and when the British tried to punish the Turkana by confiscating cattle, an estimated 14 per cent of the population died. The British at last saw promise in Turkana though and, appropriating the hostility and seclusion, the region was adopted as a prison and physical quarantine: a parched, sand-blasted Siberia.

When Idi Amin, then a lieutenant, ordered soldiers across the border from Uganda in 1962 to torture and massacre Turkana men suspected of cattle rustling, his actions were described by Britain, in the understatement typical of the era, as 'over-zealous', and if that's how you describe burying people alive, as his soldiers were rumoured to, then I don't know what cruelty looks like. Around this time humanitarians arrived, airdropping beans and making forays with medicine and food when the rain clouds failed to muster and livestock died. Catholic missionaries set up famine relief camps and treated sick, hungry Turkana who were often referred to as *maskini*, Swahili for 'poor' or 'beggar'. Among the Turkana, suspicion soon grew of outsiders' attempts to solve their problems, especially since being confined to apportioned territory, instead of being allowed to roam large distances, increased their risk of running out of food in the first place. Turkana was viewed as an island, both from without and within. Men and women in Turkana are often heard asking bus drivers leaving the region not for a ticket to Nairobi, but 'to Kenya'.

*

The track improved at times and I made gusts of progress, but the horizon still spread before me, deep and lifeless. Just as I was pondering whether I should go east to the lake shore, which could entail a brief and gory squabble with crocodiles, a few huts appeared: a mission, I saw now, as a priest ambled out to greet me, with sweat-shiny pate and spectacles.

'You're through the worst bit,' he said, as I laid down my bike.

'You mean the trail gets better?'

He looked unsure.

'I mean the fighting. With the Dassenach. We lost sixty lives here last year.'

'*Sixty?*' I squinted around at the huts, their walls made up of old US food-aid boxes, dung and plastic sheeting. Sixty looked like half the village.

Such bursts of violence rarely made the news. Conflict in these tribal lands was long-standing and assumed by many to be symptomatic of backwardness. Yet technology and outside actors had often intensified the fighting: Russian-made guns had washed in as war raged in neighbouring Sudan. As the climate has grown ever hotter, droughts have become more frequent and the new Gilgel Gibe III dam on the Omo River, allowing foreign multinationals to plant biofuels and cash crops, has led to dramatic water depletion and a lack of food. In response, tribes raid cattle from one another, retaliation employs those AK47s.

The next day the priest pointed me down a track through the Mlangoni Gorge, though 'track' was a kind description for this dry river bed, scattered with stones, snaking into the bloody Lapurr mountains. In the afternoon I noticed a few figures high on the gorge rim above, standing sentinel and silhouetted with sticks. They were looking down on me, then following at a jog. I picked up my pace, bumping recklessly over the rocks.

*

At the turn of the twentieth century, two British majors and their expedition party were slogging away from the western shore of Lake Turkana too. Major Herbert Henry Austin and Major Bright had been tasked with securing British claims to the region, which was appealingly flat and full of elephants. They set out from Khartoum at the end of 1900 with 32 Jehadia drivers (former Dervishes), twenty Sudanese soldiers, four servants, 125 donkeys, twelve mules, fifteen camels, about three months of food (rice, lentils and wheat flour) and 'a small supply of wines as medical comforts in case of sickness, tots of whiskey and a few bottles of brandy, port and champagne'. So just the essentials.

They struggled early on with swarms of mosquitoes and water shortages, crossing plains of sun-stiffened mud with cracks big enough for a donkey's hoof (the animals 'gracefully collapsed' according to Austin, who ordered them dug out with bayonets). They soon reached another plain, this one covered with locusts, about which Austin writes: 'They adopted the unpleasant practice of jumping into our tea … however the locusts were excellent eating, which was some consolation for the wrongs suffered.'

I imagine Austin shaking a filthy fist in the air: 'The brutes!' And then casually popping a locust into his mouth and chomping. 'Boys, these are quite wonderful!'

But even with these small interferences, by April 1901 the men had sighted Mount Nakua, which they knew to rise near Lake Turkana (Lake Rudolf, as it was known then). All they had to do was get there, reinforce British claims (Union Jack bunting perhaps) and march home for a pipe, 'hurrah!' and a spiffing cup of tea.

They had a problem brewing, however: the expedition's food was running out. Austin had planned to bargain for some from tribespeople around the Omo River, trading it for beads, tobacco, metal wire and cloth, but the people scarpered when the expedition party got close – memories were still fresh of brutal military expeditions from Ethiopia led by foreign white men. Austin was initially unwilling to raid villages (as previous explorers had done) and decided to move south towards the hills in today's Uganda, 300 miles away, rather than risk the chilly Ethiopian highlands, which may have yielded food but at the cost of their animals. But as they edged away from the lake and into desert, Turkana warriors stalked the party and some of Austin's men began to disappear. When the expedition party set up camp, spears were hurled in, or a ragged scream would rise up and the men would run towards the sound, rifles ready, to find one of their number speared like a table football player, no sign of their ambushers.

As the expedition progressed the camels grew thinner and were butchered. Of the men, the Jehadia drivers suffered most – they had been the skinniest to begin with. Some became hollow-eyed and reedy, while others, deficient in essential nutrients, swelled grotesquely. Desperate, the remaining Jehadia covertly slayed some of the expedition animals and picnicked on the raw flesh. Austin had one recurrent offender blindfolded, tied to a tree and shot. He docked wages from others, but to little effect since, by now, his men were sceptical that they'd survive to collect them anyway.

The Turkana warriors kept up the attacks, pouncing from the scrub ('human-tigers' Austin called them). Austin himself developed purple spots over his legs and began to bleed from his mouth and nose: the hallmarks of scurvy. A retinal haemorrhage partially blinded him in one eye. The expedition was now shambling and hopeless, fuelled by donkey flesh. Finally, seventeen of the original 72 men were rescued by Harold Hyde-Baker, a Brit stationed on Lake Baringo. The rest had perished in the sand.

*

Although I didn't have scurvy, a puncture, out here, was a portentous turn of events too. Looking up, I could still see figures on the cliffs above me, half a dozen now, watchful. It was getting late and shadows were scaling the gorge.

I set my bike down and got to work liberating Ol' Patchy, my long-suffering inner tube. Ol' Patchy was old and patchy. He'd endured maybe twenty repair jobs, but lately, in my solitude, throwing him away had become unbearable, which was almost certainly why I had another puncture now. 'There you go, Ol' Patchy,' I said, lovingly pressing the patch home and feeding him into the tyre. I crammed my underwear back inside my tyre,* pumped up the tube and set off.

Howling warriors, vipers, parboiled missionaries: such scare stories fed Ian Hibell's fears as he approached Turkana on a loaded touring bicycle in the late 1970s. Hibell had long been drawn to places judged inaccessible by bicycle, most famously lugging his bike through the dangerous, matted jungle of the Darién Gap in South America. Sucking on a pebble to stave off thirst, Hibell carried a more conservative volume of water than I'd opted for (he was a famous minimalist), ten litres sloshed in goat skins hanging from his bike. Not long after crossing the Omo River, he spotted a band of warriors with spears on a rise, about forty metres away. His heart quickened. As he tried to discern their number … three, no four, five … they charged. He

* In Sudan, I'd had a string of flats caused by a split at the base of the inner-tube valves. I figured I'd been underinflating the tubes in my fat tyres. But if I pumped the cheap tubes up towards the correct pressure, they tended to burst. My solution had been to reduce the interior space of the tyre, which I accomplished with my socks and underwear. This solved the problem, but meant that I'd been riding commando since Khartoum.

pedalled madly, but they were outsprinting him over the sand and were soon so close that he could hear them breathing. He hoped a stretch of downhill would save him, but with almost comically bad timing, his bike chain fell off and he winced, free-wheeling now but far too sluggishly to escape, like the doomed getaway of a bad dream. To his amazement, the agony never came. Instead, he noticed a Turkana warrior running alongside his bike, not looking at him at all, but focused straight ahead. After drawing past, the warrior pitched his spear, at nothing; it arced for a time and fell to the ground. The men were racing him to prove their fitness. The cruelty of foreigners had not been forgotten, but by then the world was reorganising and connecting up, and the mood in Turkana was shifting too. Hibell would end up cycling a distance equivalent to more than ten times around the equator. He died on his bike, knocked down by a driver on a Greek highway in 2008.

*

As I climbed from the gorge to Lokitaung, I clung to Hibell's more encouraging story of the Turkana and didn't dwell on the bygone violence of Austin's. The shops looked long closed, and a truck without wheels listed on its axles, picked of its parts like the dead goat beside it. It was in the prison in Lokitaung that Jomo Kenyatta, the founder of modern Kenya and the nation's first president, was locked up by the British for protesting colonial rule, a savvy choice of location to which to deport a dissident. Escape your jail cell out here and I'd imagine that you might sheepishly come back.

Lodwar is the only decent town in Turkana, and it's worth noting that Turkana is the size of the Republic of Ireland. The town has grown from famines and droughts when nomads, having lost too many cattle to survive, sought refuge. Amid shops and eating houses though, the wildness of its situation is seldom in doubt, and for a time the shops declined bank notes for tender because they could get ravaged by termites. Palls of dust, 'Lodwar rain', troubled the streets, blunting figures I recognised only at close range – NGO types, Turkana women in layers of vivid necklaces, and a few young Kenyans in their own tribal get-up: premier league football shirts. Of the 350,000 people living in Turkana, most were nomadic or semi-nomadic, bunched discretely in the desert. Some lived in extremis, slitting the necks of their goats for blood, which they mixed with milk and drank, and stitching the animals back up again so that they would survive. Others kept enough cattle to get by, and some gathered nuts and berries, hunted, traded, peddled charcoal, collected food aid, fished, and even cultivated a little when the desert allowed. It was wise to have options out here – flex or die.

I rested in Lodwar for a day and then rode 50 miles south to a village with a few wasted bushes, a dry stream bed, and twenty or so huts, walled in saplings and dung. There was a white four-by-four too, so I knew that this was the right place. Nurses from Merlin were here, an NGO tasked with delivering medical aid in emergencies. After sending a flurry of emails from an internet café in Addis, I'd been invited to visit a mobile medical clinic. I'd been on the road for almost a year now, and I longed to explore more than the terrain of the world. Perhaps, in visiting hospitals and clinics, I could think more carefully about the landscape of health, how it was moulded and eroded, along the way. A new adventure. Paul Theroux understood this drive for closer scrutiny: 'I don't see how it's possible to get to the truth of a country without seeing its underside, its hinterland, its everyday life.'

I found what looked like the right hut, a few dusty-looking kids playing outside. Inside, around fifty Turkana women were squatting on the earthen floor, wearing necklaces of yellow, red, green and black beads, holding infants to their breasts. As they turned to watch me, I hesitated to move further inside, sensing something uneasy and suspenseful in the air. Many Turkana distrusted outsiders, and plenty of suspicion was dealt to them in turn. For decades, strangers of many species had come to Turkana, many to preach, instruct and impose. Even now, the people of Turkana were occasionally patronised or debased in the Kenyan press. Historically, medics had shown little regard for their beliefs about health, and only ten doctors had ever graduated from the region. Most of those hadn't stayed around; they'd been lured to Nairobi, where the hospitals paid more, and where posts freed up when other physicians followed the money and went to work in Germany, Canada or the NHS.

Two nurses stood in the centre of the hut and now one beckoned me over. 'Stephen? Welcome. Come in, come in. We're just weighing the babies.' A blue cotton sack sagged, attached to a scale, and she picked an infant from it, tiny, head limping forward, arms and legs like tentacles. He gave a flimsy cry.

'Severe malnutrition. We'll take him to the Stabilisation Unit in Lodwar.'

The hospital in Lodwar was the only place in Turkana with the technical facilities to treat the seriously sick – if you could operate them. Like much of its medical equipment, the hospital's anaesthetic machine had been donated and I'd heard that its digital menus were in Spanish so only a few doctors knew how it worked. I asked how many were going to Lodwar. He was the fifth so far. And then the first note of despair as the nurse looked around the hut. 'We've just started,' she said.

Working in the emergency department in London, I was not generally encouraged to dwell on why my patients were sick. It was considered unmanageable, muddy, even trifling: it didn't help achieve the four-hour waiting-time target, plus it was rarely simple – people get sick for all sorts of scrambled reasons, including pure bad luck. For Africa however, the over-simplified version anyway, the Africa of high-street charity collectors and guilt-tripping posters with starving kids and western heroes, the cause of sickness was poverty. It struck me as reductive now, pronouncing money a panacea, and my thoughts strayed to the roots of poverty, and to the roots of the roots.

Health in Turkana was not governed simply by money, but by land and climate, by culture, conflict, migration, history and politics; it hinged on everything from health costs to the movements of sandflies. In roundabout ways, health was swayed by bilateral trade deals and some politician's father and the wetness of the wet season and a road built with Chinese money and the price of oil in Saudi Arabia and the Turkana's own understanding and beliefs around health and disease. It was shaped by the legacy of colonialism and the wrecking balls of poor nutrition, social neglect, the HIV pandemic, and more. It made me think of Virchow's declaration from 150 years ago: 'If medicine is to fulfil her great promise, she must enter the political and social life.'

This was radical stuff, at the time. Back in Virchow's and Austin's day, it was convenient to consider medicine capable, on its own, of solving illness; the social and environmental context was generally set aside, certainly in matters of global health, when Britain was an unrivalled superpower. Medicine in this period was often used as a defence for colonial rule: it established authority, served as a proxy for progress and civilisation, whereas disease suggested native backwardness and implied an innate, European physical advantage. It is perhaps cynical to suppose that the study and the breakthroughs in the science of tropical medicine came about first and foremost to protect

British soldiers and colonial interests, but then again, disease *is* an obstacle to profit. It's certainly true that treatments and vaccines developed in the colonial era have saved millions of lives, British and foreign, though it's also true that during previous centuries colonialists brought widespread death – the virgin soil epidemics of the Pacific islands, and the decimation of the Aztecs and of the Australian Aborigines among the catastrophes.

Malnourished nomads in desert huts – an extreme of social disadvantage, to put it lightly – but I'd seen health compromised for the same broad reasons before, devastated at times too. In London, I'd begun to spot patterns as my shifts in the emergency department rolled by. More often than felt fair, or even possible, my patients were on the edge in one respect or another; they were not a neat cross-section of Britain. Disproportionately, they lived in the poorer parts of town. Often, they were short-termist and friendless; a large number were disabled or mentally ill. The risk factors for any given disease, which I'd had by rote since medical school, and imagined as somehow autonomous, were suddenly tangled up horribly in the circumstances of people's lives. And they hunted in packs. On my penultimate shift at St Thomas', I'd treated two homeless men, one cold and suicidal, one with an abscess from injecting heroin into his groin; an autistic man with tension headaches and no close family; an obese, disabled teenager, unable to find work or claim enough benefits, suffering complications of diabetes; a poor, chronically stressed mother of four with chest pain; and an isolated Vietnamese man withdrawing from alcohol. I'd felt little sense of surprise at the time. It wasn't an unusual shift.

I had begun to feel useless, faced with the connections between social order and health, day after day. No doubt cause and effect were entangled, but still, I began to doubt the authority of medicine: how much could we do in the face of these gradients and hierarchies and constraints on choice? Even accidents looked less random than I'd assumed, and had social predilections (ask anyone who's worked in a burns unit). Like the people of Turkana, many of my patients (the homeless and addicted in particular) were taken for feckless or for victims, unwilling or unable to help themselves, and it made me bristle that agency and victimhood were seen, too often, as incompatible things.

I left the village the following afternoon. Soon, trails converged, better roads appeared and my tyres began to hum again. A chain of telegraph poles. An internet café. A car booming an American pop song. A six-foot-tall billboard advertising toothpaste. Nairobi.

*

'Missed you, mate,' Ny went in for a hug '... but you need a shower. You hum.' I grinned, glad to think of a shower at last, and glad to be reunited a month after we'd parted company.

It rained every day on the way to Kampala, and even hailed as we crossed the equator, but afterwards there was a brightness to the land; the tea was greener, the flame trees more ablaze. And then we were caught up in a city of potholes and puddles and racing *boda bodas*, motorcycle taxis with 'born lucky' signs on the front, which went some way to explain why their drivers drove without helmets, fatalistically fast. On average, five 'lucky' *boda boda* drivers died each day in the city.

It's easy to feel seized by Kampala, by its pleasant unruliness, neatly manifested at this point by two boys on a motorbike. 'Welcome to Uganda, Wayne Rooney!' On a whim, Ny had shaved her dreads to a crew cut, and now she laughed politely and swore under her breath as Kampala decided her lookalike. We wandered the city's back alleys, where women roasted goat over disused railway lines. A man handed me a leaflet. With his help, I could pass exams and get promoted. If I suffered from cancer or HIV, I was in good hands. I was promised an erection. On the tussling main streets, men sold phones, and newspapers with headlines to match the madness of the city. The inside pages were a concoction of premier league football, politics and scandal, sometimes with homophobic undertones. I picked up one paper whose inside page carried a story about a pastor accused of sexual misconduct. The headline read, 'Pastor Kiweweesi in bum sex scandal'; and the picture caption, 'Boy Drags Flashy Man of God To Police For Terrorising His Buttocks With Monster Whopper'. There were photos of the two men involved, one tagged 'accused', the other 'shafted'.

We were adding hundreds of miles to our journey by looping through Uganda and Rwanda, but we'd trusted a Belgian cyclist who'd insisted that it was worth the detour. And it was, especially towards Fort Portal when the road narrowed and the Rwenzori Mountains rose in the distance, and children ran laughing beside us. Boys pushed oversized Black Mamba bicycles loaded with green plantains, misty papyrus swamps came and went, and one morning a rustle in the bushes produced an elephant. We stalled, ten metres between us, preparing to ditch our bikes and run if she charged. But the standoff didn't last long, and she lumbered off as another rustle produced two calves, trotting to keep up.

In Tanzania, the road flattened and the rains began. Trucks soaked us in spray and whatever propelled us now was getting hard for the Tanzanians to work out. 'Is this voluntary?' they wondered. 'Is this for a prize?'

There was no need to explain ourselves on the shores of Lake Malawi though – there can be few places more soothing for a travelling cyclist. The sun drifted free of the clouds at last, traffic dropped away, and the road took in cassava crops, boreholes and women singing to babies in slings. Life felt easy now, a feeling enhanced by all the campsites and the small resort towns like Nkhata Bay, where we found light-hearted Malawian men nicknamed Laser, Fortune, Lucky Coconut, Chicken & peas, Happy and Mr Spanner, drank rum from 30 ml sachets and danced late into the night.

Our map was nearly redundant now: the Great East Road would take us from Malawi to Lusaka in Zambia, no turnings until then, so we could switch our brains off for almost five hundred miles. Amid a repetitive, rolling land, spotted with villages of thatched huts, Nyomi and I yearned for a little distance apart. We'd been together for most of the last eight months and we squabbled from time to time. Ny, always cheap in the extreme, had now turned to foraging.

'What's this, Ny?' I pointed out the green bits in my rice.

'Local herbs,' she said.

'And where did you ...?

'I pulled them up from over there,' she signalled a few straggly, anonymous weeds. 'They smell nice.'

'Ny ...'

She shrugged, apparently unconcerned about their potential to poison us. In a heated moment not long after that, I suggested that if we were scavenging, we could use her ukulele for kindling.

It never came to that. East of Livingstone the road split, as if the universe was making a suggestion. We hugged, arranged a rendezvous, and I watched Ny pedal off towards the Caprivi Strip. I took a left and began to loop through Botswana alone.

It would be easy enough to find some restorative solitude in a country the size of France with a population of just two million people, though I hadn't yet considered its other inhabitants. A sign down the road reminded me, cautioning 'Beware of the animals', but failing to specify whether that was the blinking, cud-chewing kind that shambled from time to time into the road and wrote off trucks, or the kind that launched from the scrub, tore out your jugular and turned you into a cautionary tale. I asked a man who was walking casually down the road, as you do, with a rifle over one shoulder and a dead vulture over the other.

'Oh yes! The wild animals here are many. Many, many, many! I never leave home without a gun. I saw lions here last week.'

There was a fluttering feeling in my guts now, a gratefulness for some new, invisible intensity to the world. It's incredible how lion-like a rock can appear if you're in the right frame of mind. And incredible how a piece of shredded truck tyre can move like a snake. *Wait, fuck!* I swerved; the puff adder surged for the roadside. The scrub erupted with other startled creatures too: there were antelopes, vervet monkeys, hornbills and warthogs. There were buffalo, ostrich and black-backed jackals. There were more puff adders and, on one occasion, I caught sight of something much larger: a monster, two and a half metres of Snouted Cobra, its slick, black and gold bands writhing into the brush.

In Maun, you could take a boat tour of the Okavango Delta, something that, since *Planet Earth* on the BBC, everyone wanted to do, though not everyone could. Botswana specialised in low-volume, high-cost tourism, and most of what was on offer was more than I could afford. Some guests shelled out my annual spend for one night in a lodge, others treated themselves to a booze cruise on a boat named *Cirrhosis of the River*.

Didn't matter: I was enjoying my cycling safari well enough, though by everyone's reckoning this was lion country so I camped in villages at night, sitting around small fires with young men as they talked about their dreams of leaving Africa for the US or Europe, feeling acutely aware of my own unmerited liberties. One night was different: I was allowed to sleep in the research facility of a crocodile farm surrounded by a game-proof electric fence. I felt far more at ease, listening to the bestial groans of blundering hippos being zapped with 5,000 volts.

I had, of course, missed the all-singing, all-dancing Nyomi, and I was happy when we met again in Otjiwarongo in Namibia, where Herero women walked the streets in horned headdresses and baggy, high-necked, ankle-length gowns with layers of petticoats, a style of clothing adopted from German colonisers. The Herero people were almost wiped out in the early twentieth century. 'No war may be conducted humanely against non-humans', said the genocidal German military commander Lothar von Trotha in 1904, a man who still had a street named after him in Otjiwarongo. Over the next four years, around 65,000 Herero people perished – starved, shot and interned in concentration camps.

From the quirky Germanic city of Swakopmund, sea-fogged and toy-dogged, we moved inland through the Namib Desert, on roads that ran in straight lines to quivering vanishing points. Choosing campsites only took a second now. Ny would simply nod to an available sand dune and we'd take shelter behind it. A tailwind flung us towards the South African border – 28 miles an hour, no pedalling involved, we flashed by the remains of hyenas, milking every minute, knowing that tomorrow the gods could be against us, ending the day beside the Orange River in the kind of happy state that's nearly hysterical. After six hours of barely trying, we were 209 kilometres closer to Cape Town.

The next day, in the Northern Cape, a sack of oranges was pushed through a car window into Nyomi's waiting arms. An hour later another car stopped and money was passed out, but I managed to thrust it back with a thank you. As we sat by the roadside, ripping into oranges, I thought about *ubuntu*, the Bantu notion of a universal bond. *Ubuntu* suggests that we belong to a greater whole, that humanity is a quality we owe to one another, and while the precise definition can vary, *ubuntu*'s central theme is perhaps best distilled down to: 'I am what I am because of who we all are'. I suspect *ubuntu* is not easily associated with Africa in many people's minds. In fact *ubuntu* itself, the very notion, reminds me of how little Africa is celebrated for kindness and cohesion. It is, still, a wide, forbidding place in the global imagination, everything lumped together: child soldiers, plagues, wildebeest migrations, enormous sunsets, modern day genocides, Mandela, Gaddafi. It has a legacy that's hardly going to be overturned by a football world cup and the family heritage of an epochal American president. Consequentially, Africa is rarely granted the nuance we allow other parts of the world. Writing this today, I see evidence everywhere: a BBC sports article about a football contract states that 'Rangers sign African pair Coulibaly and Umar' – it's not easy to imagine 'European' or 'Asian' in the same context. South Africa was a vantage point of sorts, and from here the continent looked more various than I'd imagined in Cairo. I suppose this is not a surprise; this grander view of things is always what travel pulls off. At home, the world is grotesquely navigable in every sense, sold to us a few digestible slices at a time, simple and processed, shrink-wrapped for our tastes.

The wild flowers of the Northern Cape turned to vineyards and the road signs no longer warned of elephants and rhinos but of tamer perils – like tortoise and golfers. Nyomi spotted Table Mountain first, a grey blip on the horizon, our finish line, at least for now. We were still gazing away, lifting our chins to the sea wind, when a car pulled up, driven by a chatty man in a suit called Paul who thought cycling the length of Africa was a fantastic thing to do. He fished out two pairs of keys and dangled them from the window.

'I've got a city house and a beach house. You're very welcome! Stay in whichever one you like.'

Paul looked at us more carefully now, taking in my fishnet-Lycra shorts, ragged vest and patchy beard, and Ny, who for some reason had a smear of bike oil over her chin. We'd spent nine months on back roads, we'd spanned a continent, and I suppose we looked at once triumphant, and like the sole survivors of a zombie apocalypse. It must have been an alarming sight.

'Guys, seriously, you must stay in both.'

I felt a nudge in the ribs. 'Let's hit the beach house first,' whispered Ny, and I agreed.

PART TWO

USHUAIA TO DEADHORSE

You remember the declension: I am a traveller,
he is a tourist, they are trippers.
A. A. Gill

4
Spine

After Nyomi returned home, I flew: Cape Town to Buenos Aires to Ushuaia, a city bullied by wind, on the edge of an island in the Tierra del Fuego archipelago, where South America fractures and leaves a cluster of islands to the everyday furies of the Southern Ocean. They say Ushuaia is the end of the world or, I supposed, the start – depends which way you're looking. For me, Ushuaia was a new beginning, the launching point for a slog between two sparse worlds, one that sounded relaxed and homely, like a middle-class dad in a fleece, the other a little frightening, like a grizzly bear eating your head: Patagonia and Alaska.

Their tendency to cold dictated my timing. Summer was merely a good idea for biking through Patagonia, but it was essential for Alaska, so I had from the beginning of one summer to the end of another, around twenty months until the weather window closed. In a few weeks, I'd be huffing through the Andes, and if that chain of mountains could be likened to a spine, Ushuaia really was the arse end of nowhere. My plane swooped low over the Southern Ocean, and whitecaps and spume filled my window as we were pushed around, wind-jiggled, passengers clutching hands by the time a runway appeared. On landing, there was snowballing applause. Stepping off the plane, I was in the most southerly city on earth, closer now to the South Pole than to Argentina's northern border with Bolivia, and you could sense Antarctica's ghostly proximity here, the wind sharp as ice, moaning around the airport.

I'd over-indulged in Cape Town; my fitness had waned and my bike felt heavier, or perhaps I was, but cycling out of Ushuaia I lapsed again into familiar rhythms, and felt happily reunited with a life spent thoughtfully picking out small details of the world. 'This capacity to wonder at trifles', said Nabokov '... these asides of the spirit, these footnotes in the volume of life are the highest forms of consciousness, and it is in this childishly speculative state of mind, so different from common-sense and its logic, that we know the world to be good.'

Childishly speculative: you and me both, I thought, as a Patagonian fox rolled around playfully in the road, almost at my feet. Later that day I spotted a beaver gliding, sleek, through a small lake and diving from view. The species is not endemic to Patagonia but was introduced from North America in the 1940s for commercial fur production, an exercise that back-fired spectacularly: within 50 years 25 randy pairs had turned to 100,000 individuals. They've been culled now, and presumably, without trying hard to aim.

There are stark and beautiful parts of Patagonia, especially when the land rears up to form the southern Andes and glaciers fill the cracks, but mostly it is a dreary, flat and whistling place, without a great deal living in it. I felt unable to wonder at trifles now. I watched shingle do nothing for days. Oil fields accentuated the tedium, their pump jacks bobbing, bobbing, and it felt like water torture. I perceived in the land not hostility exactly, but a mean-spirited indifference, and each hour spent here felt too long. I passed the time conjugating Spanish verbs and reciting all the causes of peripheral neuropathy. I tried to recall all the places I'd camped in Malawi. I dwelled on unlikely scenarios involving beautiful and lonely Chilean women, bored on their farmsteads, feverish with lust. I missed people.

Miles are never more tedious than in a headwind. It's not just sailors and airline pilots for whom the wind is a vital current, something to be respected and reckoned with, tactically outmanoeuvred if you can. Long-distance cyclists too are subject to the quirks of air flux and some plan their routes with prevailing wind directions in mind, though in Patagonia the winds tend to be too flighty to predict. I was here as a teenager with my brother, and I remembered how the wind pinned us back then too. One morning, we'd emptied our tent, unpegged it and then watched numbly as it was snatched away. Its maiden voyage lasted two hundred metres, a hooping flight across the pampas. We caught it, but thorns had torn us a skylight.

Now, walloped by the wind again, I ground out the miles, my head bowed in deference. One morning, I looked up to see the granite fang of Mount FitzRoy, and closer in, shadows of condors were dashing over the lower hills. For the next few days I pushed through thin forest trails, wading rivers and trudging over marshland, shouldering my bike over deadwood, all for a weekly ferry that took me to Villa O'Higgins, the southern end of a road that would carry me for hundreds of miles north: the Carretera Austral.

The Carretera is a choppy ride that takes in forests, fjords, milky rivers, glaciers and the stocky mountains of southern Chile, stretching for over a thousand kilometres all told, sometimes paved, but much of its tail end still offering the satisfying crunch of gravel under wheel. Summer brings a great migration of touring cyclists to the Carretera, northern Europeans mostly, and at times I could barely move for Belgians. We met headlong or sidled up to each other in lay-bys and some days I found myself in a peloton of bikers, ten strong, that swallowed more solo riders and pairs. It was a happily social time. Only the *tabano*, a biting breed of horsefly that tracked us up the hills, spoilt the party, feeding on us in our weakest moments.

They're unkillable. Think you've squished one and they re-inflate, buzz away and bite your eyelid.

I spent Christmas just over the border in Argentina, and, from a guesthouse in Bariloche, I peered out the window every now and then back towards Chile. I was staking out The Cloud. The Cloud could have been some lonely, benign clump of cumulus, but the residents of Bariloche, glancing anxiously in its direction, knew better. Several months ago, after lying dormant for half a century, a volcano named Puyehue erupted, not from its old caldera but through a new gash in the earth's crust, six miles long by three miles wide. Bariloche, two and a half miles away, had been covered in a thick coat of ash, and as the earth continued to belch out thick plumes, airports had to be closed on the far side of the Pacific, in Melbourne.

To the north, in my path, the ash-cloud loitered. On the venerable Route of the Seven Lakes, I counted only two; a grim haze had erased the rest. Sodden ash smothered every tree and mountain for miles. The sun was like a torchlight weakened by fog, but wind didn't lift the murk – the air was too thick and burnt. Lorries sprayed the streets with water, cars drove with headlights on full beam. There were post-apocalyptic flashes of people shuffling along, holding handkerchiefs to mouths or wearing surgical masks. My own mouth dried up and my eyes watered so much that I could see the red–amber–green volcano warning systems outside villages only when I was right in front of them. I met just one biker, a dimly seen, Mad-Maxian character, who emerged from swirling dust in swimming goggles and then moved off again, unspeaking, a ship in the night.

Roads meander out of the Andes on both the Chilean and Argentinian sides like nerve roots from a spine. Deep within the spinal cord were the high passes, and I was hooked from my very first crossing between countries, on Paso Mamuil Malal, which twitched through monkey puzzle forests beside the stark cone of Lanín, a stratovolcano. It was the start of a long addiction that would last the length of South America. I was tempted back again and again, to Paso Vergara, Paso Pircas Negras, Paso de San Francisco, digging my bike wheels out of surreal, black waves of basalt, sucking up sulphurous air, crossing the border between Chile and Argentina ten times, trending north, and sleeping in *refugios* when I could, or camping in the Puna, a region of high grassland that lies above the treeline. With each pass, I felt more capable of the next. I got hopelessly lost, a few times dangerously so.* But I never felt that the struggle was in vain because no mountain range has the colours of the Andes, where salmon, cream and peach can swirl together, like the blending gases in artist impressions of distant planets. I felt more physically present when I was weathered and overtired and a little hungry for air. I began to notice a hyper-sensitivity to my body in such moments, to its small physiological details, to steady breaths and tickling beads of sweat. I felt simultaneously miniaturised by the mountains, and part of something larger. When Proust wrote of emotions, he couched them in physical terms – 'geologic upheavals of thought'. I could see why.

I felt a long way from my old life, back when I was part of the millions huddled in the city and indoors. Nature, back then, came occasionally in bursts, on short holidays, or in bed, with a David Attenborough narration. I'd felt detached from the natural world, which should be an odd thought really, a bit like moaning at being stuck in traffic when you *are* the traffic. Around the time I left home, a literary trend was emerging of nature as cure, and, taken together with the term 'nature-deficit disorder', coined by Richard Louv in 2005, as well as the

* My sense of direction is exceptional and I'm sorry to say that I still mutter 'Naughty Elephants Squirt Water' in order to discern east from west.

countless eco-therapies on offer for city dwellers, it sometimes felt as though wilderness was becoming medicalised, as if people were dosing up on nature, taking hits to tide them over. Now, amid the mountains, I wondered why this concept of nature felt so misguided. Perhaps because it seemed to suggest that the consequences of wilderness were predictable. For me at least, it was the intense *unpredictability* of being outside in wild places that I treasured most, the various erratic, overlapping sensations I was left with: fear, despair, transcendence, solitude, weariness and awe. Upheavals of thought.

The most famous of the Andean passages was Paso Los Libertadores, the busiest of the forty or so connections between Chile and Argentina. The road clung to Río Juncalillo at first, but when the bond was broken it was a dramatic parting. The river idled away around a corner as the road took a desperate leap up the side of a mountain in a series of sharp, intimidating chicanes. Locals referred to this climb as Paso Los Caracoles – Snail's Pass – since everything climbs at a snail's pace. Hours later, the river became part of the allusion too, a sinuous glimmer, like the trail of a snail on a winter morning.

In the afternoon a Chilean biker, riding down, stopped for me in the road and we talked a little. His father had died six years ago, and his ashes were scattered at the top of the pass next to the four-tonne statue of Christ the Redeemer, on the old road that marks the frontier. On this day, every year since his father's death, the biker had cycled Paso Los Libertadores. Every year he talked to the dead man as he pedalled. 'Sé que puede escucharme', he told me, dropping his eyes as they filmed over with tears. *I know he can hear me.*

Christ the Redeemer of the Andes was donated by the Bishop of Cuyo in 1904 to help ease tensions between Chile and Argentina, who had been on the verge of war. The statue was carried up piecemeal by mules and the two armies fired gun salutes together instead. A plaque on the statue reads: 'Sooner shall these mountain crags crumble to dust than Chile and Argentina shall break this peace which at the feet of Christ the Redeemer they have sworn to maintain.'

Or, I think this is what it says. I was squinting through a drumming headache, attempting to translate. Climbing 2,600 metres in a day on a heavy bicycle is a good recipe for altitude sickness, but particularly when you're me. There is something wrong with my physiology. On every occasion I've been at altitude with other people, I'm always the first to get sick. The oxygen falls by the most minute fraction and I'm a cussing, vertiginous invalid, holding my head in my hands and panting like someone who's been told they need to perform eye surgery on themselves with a spanner. With no desire to camp breathlessly with Jesus, I rushed downhill, feeling better with every mile, the atmosphere gradually crushing my plastic water bottle instead of my head.

Back in the nineteenth century, as adventurers began clinging onto icy crags and toying with balloons, nobody was quite sure why people got sick at high elevations. Some thought it was rupturing of blood vessels, the intensity of the light at altitude, poisonous mountain gases or even atmospheric electricity. The early Victorian balloonists, immensely stoical men, of course, had some pretty messy experiences with altitude sickness. Glaisher and Coxwell, a British meteorologist and a dentist, took off on 5 September 1862, and in what we now recognise as a crap plan, ascended to 8,000 metres in 48 minutes. Glaisher reported that his arms became powerless and that he could not speak. 'I tried to shake myself and succeeded, but I seemed to have no limbs.' The men pulled a cord with their teeth to open the balloon's valve and managed to descend. They barely survived.

Other balloonists did not. The most famous disaster,

in 1875, was that of the balloon *Zénith* and its three French adventurers who carried oxygen-air mix in an attempt to break Glashier's altitude record. Tissandier recalls feeling inner joy as they ascended to over 7,000 metres before there was a sudden feeling of weakness. The men slipped in and out of consciousness and when Tissandier at last woke up the balloon was descending rapidly and both of his friends were bloody-mouthed, purple and quite dead. Newspapers declared the men martyrs to science and, at their funeral, it was eulogised that, 'our unhappy friends have had this strange privilege, this fatal honour, of being the first to die in the heavens'.*

*

Released from the heavens into the north of Argentina, I stopped for a few days near Salta, a town surrounded by attractive *quebradas*, gorges, which I planned to explore over a few days. In a campsite, the owner came over to say hello and shake my hand. Mine was lost in his sweaty, doughy palm. I looked up. He cut a large figure. His face was an oval, his nose bulky too. He had a heavy brow and a lantern jaw. And then he spoke, a low, gravelly drawl, and I thought: *I know this*.

I didn't see him for the rest of the day, but later, when I saw his wife, I plucked up the courage to ask a question that could have sounded both intrusive and nuts.

'Sorry, does your husband have any medical problems?'

She smiled. He'd been diagnosed with acromegaly two years ago.

It's a condition caused by excessive growth hormone arising from a tumour, generally benign, in the pituitary gland, which

* Quote is from http://www.thosemagnificentmen.co.uk/balloons/zenith.html

lies deep in the brain. Surgeons often remove the tumour, going via the nose to do so. He'd been operated on too, but many of the features of acromegaly remain. That I met him at all was chance: you will find only sixty acromegalics in a million people. As the tumour pumps out growth hormone there begins an insidious transformation that can go unnoticed for years. Most people think it odd, not indicative of a medical condition, when they outgrow wedding rings and shoes, when their teeth grow spaced out and they begin to snore at night. When it was first described, doctors knew nothing of hormones, and acromegaly was seen as a disease of the bones.

There are some acromegalics you will have seen and heard of, as the drama of their appearance has opened doors in acting roles: Carel Struycken, who played Lurch in The Addams Family; and Richard Kiel, who played the character Jaws, the James Bond villain. Their size too is suited to a career in wrestling: Big Show and André the Giant had acromegaly. All the world's tallest people had acromegaly in childhood, prior to the bone plates fusing: in this case, it's termed gigantism and is rarer still. Acromegaly is a cruel, deforming disease, though it has been treated less sympathetically in the past. Mary Ann Bevan was a poor British woman with the condition who became a star of American sideshows in the 1920s, hired as 'the ugliest woman in the world'.

As a fresh-faced medical student I would meet acromegalics as part of clinical exams, and, while the disease was another shocking example of how the body could fail, it was also a helpful reminder that, however patients described them, their problems were inscribed on their bodies, too. This is perhaps the mode of practising medicine I enjoy most: doctor as code-breaker, listening for clues, tracing constellations of symptoms and signs. With experience though, you come to see grey zones in diagnoses, or multiple problems at once, or realise that what matters most is how the problem you've diagnosed relates to the patient, or the patient to the problem, because we're all individuals, in life and disease.

But to me, in those early, quixotic days of medical school, doctors seemed more than mere detectives; they were sorcerers, could read fortunes. The wild promise of phrenology and palm reading have nothing on the predictions you can make if you know the puzzles of clinical medicine. Take, as one example, what you can deduct from a seemingly unimportant part of the anatomy: the fingernails. They can curve upwards (*koilonychia*, suggesting anaemia), turn white, yellow, blue or green (with disease or for no comprehensible reason at all), or be half one colour and half another (inexplicably common in patients with kidney failure on dialysis). They can be suggestively ridged and pitted and lined, or dashed with 'splinter haemorrhages' if there's an infection in remote anatomy: the heart valves. They can be 'clubbed', once known as Hippocratic Fingers, a convexity of the nail and sometimes a thickening of the fingertips. 'Oooh, check out those chicken drumsticks' whispered one consultant, leering over my shoulder as I examined a patient's clubbed fingers. My textbook listed thirty causes of clubbed nails, and many were afflictions of the heart, gut and lungs.

There's more: a fingernail can be a record of a life, like tree rings. Beau's lines are deep horizontal grooves caused by a short period in which the nails stop growing, usually because of a severe illness or chemotherapy. Otzi the iceman, the oldest-known natural human mummy, discovered in the Alps, was found to have three Beau's lines on a fingernail, which suggested that he'd had three major illnesses in the six months before he died. What an odd universe the body can be.

And that's just the nails. Move up and things don't get any less fascinating. The pulse, I'd discovered, was not merely a throb under my fingertips, it had personality: jerky or 'waterhammer', bounding or slow-rising – all fragments of the story and clues to diseases of the heart. For a tremor in the hands, the devil was in the detail: was it fine or coarse, and at how many hertz? Was it a sign of syphilis, Parkinson's disease or forty years of Special Brew for breakfast? And there are signs everywhere: a small diagonal crease in the earlobe (Frank's sign), has been linked to a higher risk of heart disease. How mind-warping is that?

*

Paso de Sico was my final venture over the border, from Argentina back into Chile. On the wall of the immigration hut, guards were marking off the days they had left like prisoners, and I wondered how the blunt beauty of these flowing, coral mountains had ever lost its thrill. Beyond the pass I dropped through the high desert to the town of San Pedro, where I found a campsite and, in it, Nicky, a lanky, stoned bike mechanic from Birmingham, pedalling through South America in an old man's flat cap. He called me 'brother', 'our kid' and 'ripper' and I liked him immediately.

The relationship I had with my bicycle was a tangled one. I loved and resented it equally and in a childish way I hated how my own mechanical skill determined its performance – there was nothing Zen about the art of bicycle maintenance for me. I'm just not particularly practical, and I imagine there was an enormous sigh of relief when my professional indemnity insurers were told that I'd decided not to be a surgeon. For a cyclist like Nicky though, the relationship had the purity of a love affair. He tinkered lustfully. They say there are cyclists who travel and travellers who cycle, and if that's true, you can tell the type by their stories. Nicky had a tale of a horrific crash he'd suffered years ago while riding back at home. A truck had run him off the road and he'd tumbled down a verge. He spoke in gory detail of the damage to his spokes and forks, looking bereft. When I asked him how he'd fared, he added, as an aside, 'I bled for a while. Lost some teeth. Broke my jaw, oh yeah, my wrist too. But man, you should have seen the front rim, it was really twisted, brother!'

Nicky and I set off together across another border, towards half-deserted Bolivian towns and a string of high, freezing mornings on the Altiplano, the greatest area of high plateau on earth outside Tibet. I pointed out the struggles ahead. 'Good job we're fucking hard then,' Nicky replied. In case we weren't, I suggested we have a day off in the small town of Uyuni, where perhaps we could find some fun. We happened upon some in the aptly named Extreme Fun Pub, where the menu listed unwholesome-sounding cocktails that combined the usually unconnected things, at least to my mind, of llamas and sex. Nicky ordered 'Orgasmo múltiple de la llama'. I dithered. It was either a 'Llama's sensual navel' or a 'Llama Sutra'.

We were in Uyuni to pedal over the salt flats of course, taking in at ground level one of the brightest marks on the earth, one that Neil Armstrong assumed was glacial ice as he pondered our place in the universe from Apollo 11. The next day we joined a gleaming white honeycomb, tiny ridges of salt marking the edges of the so-called salt tiles. The ground beneath my wheels was firm but felt frangible, as if I were riding on ice. Without roads, we simply headed across.

After a while, we ran into a shallow covering of water, overflow from nearby Lake Poopó, faultlessly reflecting the sky above. Nicky rode ahead, skimming over clouds. There are few square miles on earth as flat as the Salar, and, inspired, we threw our arms in the air, closed our eyes and wobbled blindly over the salt. We camped, but it was hard to sleep. The cold was biting and with the salt lit by a full moon, the temptation to take another peek at the ghostly expanse outside was too much to resist.

After circling lakes red with algae, after riding more corrugated gravel roads, and after occasional evenings on remote mountainsides, over 4,000 metres high, crammed inside a tent watching *Only Fools and Horses* on Nicky's laptop, we said goodbye in La Paz. He'd met a girl and had now dedicated his life to keeping up with her as she travelled through South America by bus. Those gruelling 200-km days were worth it – she returned with him to Birmingham. They had a kid, last I heard.

A few weeks later I watched grey oblongs drifting along, fusing into longer shapes, splitting again. The Peruvian coastal highway looked no fun at all. Trucks drifted out of *la Camanchaca*, a dense sea fog that invades the coastal desert on the back of an onshore breeze, drifting over sixty miles inland. Towns were depressingly out of season. I passed a plague of empty *restaurantes turísticos*, deserted amusement parks, and dilapidated hotels offering bad ceviche and a soft drink.* The Pacific Ocean was a murky green with a white ribbon of froth and fizzing waves. As I watched turkey vultures gather around a washed-up seal carcass, feasting, I felt a pang for the mountains. The only boon of Peru's coast was speedier progress, but it didn't feel enough when there were Z-shaped scars on mountainsides to ride. I hung a right and craned my neck again.

Hills came first, iridescent with drying chillies. The back roads got slimmer, villages further apart. Colours dried up, and only the jacaranda stood out, violet by day, a soothing purple by evening. The Marañón River criss-crossed my journey north, so I'd drop two or three thousand metres and

---

* Fun fact: a 7up is a 7up in any language – 'siete arriba' will get you nowhere.

then climb the same, rinse and repeat. Sometimes I climbed for an entire day and by dusk I could still look down upon the spot where I'd had lunch, the pueblo where I bartered for mangoes, even the field where I'd camped the previous night. I was fitter than I'd ever been and in a certain light the pattern of veins on my calves had a creepy likeness to Che Guevara's face.

One evening, the sky swiftly darkened, a belt of rain clouds blending with dusk. I spotted a house away from the road, an aloof, box-like home, with no glass in the windows and weeds bursting from the cement and brickwork. Derelict, I assumed, setting up camp under the tin roof, which jutted out beyond the outside walls. My tent was not entirely waterproof now and this was welcome cover from the rain.

I woke up, unsure if I'd heard something or if I'd been dreaming. I lay still, listening, the sounds of the night washing in.

Crunch.

Footfall? I waited for another. Only a prickle of rain on my tent now, and the wind, tugging softly.

Crunch crunch ... crunch

Fuck.

The footsteps were close and precise. Someone was circling my tent. I was being surveyed.

The blue glow of my watch said 3 a.m. I sat up, unzipped my tent and peered into darkness. Nothing at first, then something, the hint of *someone*. The figure took a few strides towards me until, suddenly, I was staring at knees. A face appeared in the porch of my tent, something in his right hand, *shit shit shit* – it looked like a gun.

When he raised the revolver, it looked illusory and weird. I thought: *breathe*. Then: *talk*.

English words came first, then Spanish in a messy flood, words clambering over themselves, pronunciation gone to shit.

'Wait, stop, I'm a traveller, it was raining, I needed ... What do you ... no necesitas el arma!'

'Fuera' – *Get out*. Not angry but not calm either. I twisted out of my sleeping bag, yanked on some shorts and scrabbled to leave my tent. I stood up and he stepped back, the revolver pointed at my guts. He was wet with rain, streaked with mud, his eyes wide and alert, uncertain ... so he was afraid, too. His gun hand was shaking.

He angled the gun up a little. One twitch of his finger and that would be it – a bullet punching through my chest, my lungs, my heart, my aorta, my spinal cord.

'Get into my house.'

There was a tremble in that voice too. My mind was snagged, but there was nothing left to do. I turned my back on the gun and walked. He followed me inside, lit a gas lamp and a room sped into view. Not derelict then, but barely lived in, just a table, two stools and a stove in the corner. The lamp sent watery shadows over the cement walls. He motioned for me to sit.

'¿Que quieres de mi?' *What do you want from me?*

I blanked out the gun, tried a slow, clear voice.

'I'm just a tourist from England. My name is Stephen. My bicycle's outside. I needed somewhere to camp.'

His eyes dropped away and he scrunched up his face, seemed to be thinking. Seconds passed.

'The rain ...', I said, but stopped, sensing that the quiet was a favourable sign. More seconds. He nodded.

'Si, si. Esta noche hace frío.'

And then '¿Quieres sopa?'

Would I like some soup?

I thought so.

'¿Pollo o tomate?'

'¿Qué?'

'Tengo pollo o tomate.'

'Tomate.'

He turned to face the back of the room, laid the gun down and fiddled by the stove. In a few minutes he returned with two bowls, steaming with tomato soup. We sat, began to talk and eat.

Asto told me that men had come to his home last month with guns. They'd taken everything. Afterwards, he'd bought his own gun for protection.

'Why are you back so late?' I asked.

'Oro,' he said. *Gold*.

Gold! It made sense of the hour, the muddy clothes, all those holes I'd noticed cut deep into the hillsides. Asto had been mining. Illegal, unless you're a licensed multinational, but local men ignored the rules and made nocturnal forays.

He dipped into his pocket and brought out a wad of tissue paper, which he opened up: two nuggets glinted in the wander of the gas light and we grinned over the night's treasure trove.

He spoke of a family, a wife and three young children in a poor industrial town on the coast. He could make better money for them up here. He'd head back soon, sell his finds.

He took my bowl. 'If you need anything, you can knock. Buenas noches.'

'Muchas gracias.' It came out loud and with feeling.

I walked back to my tent. The rain had stopped and a few stars were out. I fell asleep again next to Asto's home, sensitive to the murmurings of the night. There was a lulling whisper to the wind now, and in a few hours the sun would rise.

More footsteps as the sun warmed my tent. A voice, 'Esteban! Esteban!'

I peered out. Asto was holding a bowl of tomato soup: my breakfast in bed.

'To give you strength for your journey. I wish you luck in my country. Please, be safe.'

I smiled, and he smiled back, and I thought that he knew what I was thinking, that we were sharing a joke, that I was trying to stay safe but then a wild-eyed stranger put a gun to my face at 3 a.m. I find it hard to eat tomato soup these days without thinking: *I nearly died once.*

*

I crossed into Ecuador at the small border post at La Balsa, further east than most other crossing points (and most travel east of here is done via river rather than road). Forest and jungle should be a rousing adventure, but it is rarely what you hope for. A hot, sticky, Planet of The Insects. It's nice to watch a blue morpho butterfly glide about, but then a Peruvian giant centipede invades your shoe and all manner of flying nasties fill your tent or bounce off your head torch into your pasta. Yep, that crunch and explosion of bitter goo was an invertebrate – best not check which one – swallow hard and get used to it. Plus, this part of the journey was marred by perpetual rain, and the roads ran with liquid mud.

There was a lot more cloud forest in Colombia to navigate before I finally reached Cartagena on the northern coast of South America. Ahead now lay the Darién Gap: wild, lawless, roadless; an indomitable jungle dividing Panama and Colombia, infamous for wandering drug smugglers who hang corpses of their enemies in trees. I felt somehow reassured by its existence, though: in a world where people have driven cars to the North Pole, here was a region that few people have the *cojones* to explore. Given that this squarely included me, I sought out a chartered yacht.

There were several on offer, but, as I'd heard rumours of

disreputable sea dogs capsizing their vessels for insurance payouts, I selected warily. The *African Queen* at least looked the part: a 40-foot catamaran captained by Rudy, a sea-tanned, curly-haired Italian, a supreme chef and occasional raconteur who swore in four languages and bragged that he sailed with more rum aboard than water. We set sail with a few other travellers and Rudy's Colombian girlfriend, 30 years his junior, who dipped long curved nails into a bag of cocaine from time to time, sniffed it up and smiled.

The boat moved along with a tail of phosphorescent algae, accompanied at times by dolphins and manta rays. We stopped only at the San Blas Islands, an autonomous region partly inhabited by the Kuna Indians (foreigners having been kicked off years ago) and – this is true – a place in which, until relatively recently, the primary unit of currency was the coconut. Finally, the low, grey mountains of Panama rose from the sea, as sinister as shark fins.

I'd spent a whole year getting to know the forests and mountains of South America, and now I needed to salvage some time before Alaska got absurdly nippy. The easiest way to do this was to stick to the coast, so I put my now athletic legs and extra haemoglobin to good use, turning Central America into a blur of beaches and coffee plantations. Generally speaking, something untoward is going on if you're changing your underwear less often than countries, but, in my defence, I was a much faster cyclist now, the roads were paved and the countries were diddy.

By El Salvador, I had taken to camping on grass beside petrol stations because they were often patrolled by armed guards who were generally happy to include me in their jurisdiction. On one such night, I woke with a feeling for which there is a specific medical term – *formication*. It is the sensation of insects or ants crawling over you, though it is generally a feeling *akin* to this, a hallucination, occasionally experienced during alcohol withdrawal, or even in Ekbom's syndrome – 'delusional parasitosis' – in which patients are convinced they are infested and interpret any mark on the skin to be proof. Such patients have been known to attend clinics with matchboxes of 'evidence' containing insects – clinicians call this 'the matchbox sign'.

Unfortunately, I did not have delusional parasitosis. It is impossible to ant-proof a tent. They leak in, through gaps at the end of zips and tiny punctures. Scouts will leave chemical trails for others to follow if they strike biscuit. I remember this particular evening with the kind of clarity a French village boy might recall the day of German occupation. In the early hours I sensed my eyelids being stung. I patted sleepily at my face and then heard a voice from outside, and recognised it as the guard's.

'Esteban, Esteban! Hormigas de fuego!'

Fuego? That's flame, or fire. I sat up. I was not on fire, good. *Hormigas?* I knew that one too. What was it? Ants? That's it. Ants Fire? FIRE ANTS!

A flailing hand found my torch. The beam illuminated a pullulating, dripping ball of fire ants on the roof of my tent, pouring inside through a half-centimetre slit. I jumped outside and watched in horror: my tent was seething. Ousted, I took refuge inside the petrol station and by morning, the plague was gone.

*

Mexico. I gaped at my map – the first Mexican state was the size of Scotland and there were eight more states to cross until the US border. Mexico, in common with Turkey and Kazakhstan, is one of those places that's far more imposing in reality than you've been led to believe. To traverse the whole country, heading north along the coast to the US, I would cycle more miles than I would on the entire US west coast, from San Diego to Canada.

I took few days off now, but I took them seriously, body-surfing in the sea until beer. I slept on beaches in hammocks, hanging out with surfers who were drifting up and down the coast, hunting swells. Canadians too, in need of sun, banned from the US for drug offences dating back decades. Quebexicans, people called them. A surprising spree of wildlife kept me entertained: iguanas scuttled across the road, ospreys glided above, and a rattlesnake on the trail gave one Wednesday afternoon an unexpected spasm of terror. Finally, I got a boat to Baja, and rode that long, thin peninsula, more miles than Land's End to John O'Groats. Dusty towns, where men sat with their T-shirts hoisted, airing their paunches like overheated cars with the hood up, came and went. Chihuahuas ran yapping at my wheels.

Then finally, Tijuana. Mexico was over but there was an encore: a man dressed as a yellow duck. His head looked out through a large bill and his feathers shivered as he danced by the road to electronica, and in such a frenzied fashion that it took me a few moments to see that he was not on a stag party, nor off to run a marathon. In his hands, there was a sign advertising a nearby pharmacy – so he was simply at work, in marketing. Heartsick, since all of Latin America was at an end, my home now for more than a year, that dancing duck seemed to capture something of the liveliness I'd confused, too often, for chaos. Out of respect for the raving man-duck, I did a little shopping in the pharmacy, then looked north where all of America was waiting.

5

Flatlines

I'm an eighties kid and this grants a particular view on the USA. The 1980s were pretty kind to America, partly because of what came before – the gloom of Watergate and Vietnam, the bleakest moments of the Cold War. Millionaires were multiplying fast, and America was still an unrivalled superpower with all the pluck and confidence that begets.

Like most people my age, I grew up in thrall to the USA. On the big screen, America in the late 1980s looked fun and venturesome to kids in particular. We went ghostbusting and back to the future. The Goonies, Bill and Ted, and all of those *Stand by Me* fellas were having a whale of a time. And then the 1990s rolled around, and I came of age in a golden era: hip-hop was an exciting pandemic. North Oxford was a far cry from the South Bronx – fewer block parties, more farmers' markets – but even so, hip-hop permeated my days and nights growing up. I was hypnotised by voices from another America, wisecracking rebels, whom I admired.

But it was only ever a long-range relationship, and it felt late, now, to be visiting for the first time in my life. To do so, I needed a visa waiver, which meant navigating an online application and a stunning set of questions that only posed more. Did the Department of Homeland Security *really* believe that someone capable of genocide would be unwilling to lie about it on a form? And why would a top-secret operative confess to 'international espionage' under the duress of an empty tick box? No, I confirmed with a click, I don't have cholera.

I left my hostel in Tijuana at dawn and wheeled my bike towards the gates of a colourless immigration building. Mexicans with rucksacks and suitcases formed a line stretching untidily into metal turnstiles, which a few roly-poly women were negotiating while children clung bashfully to their legs. A space appeared in front of me and I was beckoned over to a desk by an immigration officer and then studied, as you might do mould on cheese. The face of the USA demanded my passport. He was a big guy, Tex-Mex for breakfast big, with an emphatic chin and half-shut eyes. He eyeballed my passport, lingered over my Syrian visa, leaned back, indulged himself, scowled at the visa, me again, back to visa, as if psychological pressure would crack my cover story. I smiled.

'Occupation?'

'Doctor.'

This is a sort of guilty pleasure for all doctors. Open sesame. Syrian holidays aside, I had the right job, the right skin tone and the right nation named on my passport, one perching near the top of the world's geopolitical hierarchy. It was a full house of privilege, and he softened.

'How long you been riding?' He considered my bicycle now.

'Three years.'

'Huh. Three years? Well now, wouldn't want you as my doctor.'

'That's a fair point.'

He nodded, as if to say: *I'll make fair points any time I like, buddy* and then looked for a moment like he was trying to smile (trapped wind?) before I was admitted into the US of A.

Typically, the first day in a new country brings some quirk of fortune that seems to guarantee good times ahead – call it the First Day Effect – though perhaps I'm more receptive then, or perhaps it's a falsity and my memory clings to these moments more than others. I was given money and oranges on entering South Africa. My first outing over the Argentinean pampas was interrupted by a farmer insisting that I join him for a drink of maté: simple, but somewhat symbolic. In San Diego, I sat outside a café called Kansas City Barbeque, remembering these episodes, waiting patiently for the miracle to happen, when, miraculously, it did.

A sign behind me on the wall had the logo for the film *Top Gun* and the words 'Sleazy bar scene filmed here July 1985'. It felt like the American equivalent of one of those signs in London saying, 'In a house on this spot ...'.

'Really? Maverick and Goose sat here!?' I asked the waitress whose name badge said 'Adra'.

'Goose, you big stuuuuuud! Take me to bed or lose me forever!', she said.

Adra had light-red hair and the habit of smiling slowly and dreamily, as if dosed with morphine. She'd worked in cafés and bars from here to the Midwest, and seemed at peace living by loose, interim plans. She walked around my bike, pinging the bungee cord, asking questions. After spieling through the places my bicycle had landed me so far, Adra said I could crash at her place for the night, and with that, a sunny waitress replaced a dour border guard as the face of the USA.

I would have taken more time to flop about in San Diego with Adra, but as ever, time was short, so the next day, intent on making America as American as possible, I cranked up the hip-hop on my iPod and cycled off saying 'Hey!' to people, recklessly, without invitation to do so. The Americans enjoyed this. This was Highway One, after all, and American roads symbolise freedom like no other roads can (even if, like this one, they were constructed by convict labour using dynamite and steam shovels). And while it didn't have the pop culture whump of

Route 66, it was still an American classic, and I was happily companioned by the wide ocean on one side and a wide country on the other – it seemed a momentous way for America to begin.

The hills that characterise parts of the Californian coast hadn't materialised yet, and it was easy, knocking out miles over this smooth tarmac, the highway ahead a long flat line. By now, I'd become a little preoccupied with squeezing miles out of my days and I tallied them up religiously in my journal every night. Every 1,000 km, I'd stop to take a photo for my blog, spelling out the distance with whatever was at hand – scribbling it into sand or mud or using fallen branches of trees. I'd treasured this growing collection of milestone photos, but now the 40,000-km shot felt senseless. Did hoarding miles reflect my obsessiveness, a personal peculiarity that had seen me through medical school and beyond? Or did it simply echo a cult of productivity to which most of Britain belongs? In *Walden*, Thoreau mused over the definition of success. 'The true harvest of my daily life is somewhat as intangible and indescribable as the tints of morning or evening. It is a little star-dust caught, a segment of the rainbow which I have clutched.' Milestone photos were not stardust. I skipped the next one and the one after, feeling lighter, unshackled, though there was a faint sense that I'd need to think harder about my intentions too. I'd been away from my job for over three years now and I missed it at times. No doubt, at such moments, one of my colleagues was splattered in three different body fluids, or gloving up for a manual evacuation of a delirious, heavily constipated 85-year-old, but, gross moments aside, as a doctor, I'd only rarely considered the why. Lately, though, I often questioned what drove me on, especially on the rougher days when I was overtired, my brain slopping around like porridge, and with nothing in particular to show for the last four hundred miles except circuitous thoughts and questions without answers.

*

I couldn't fathom where Los Angeles began, but apparently, I had been assimilated. A billboard advertised a 'Medical Marijuana Doctor' and on the accompanying photograph a deeply tanned, peroxide blonde ('Laura: our patient coordinator') smiled impishly, as if she knew just how wonderfully subversive and pseudoscientific California's medical-use marijuana laws were.

I joined a bike lane by the dry concrete bed of the LA River. A lane for bikes! It felt like years since I'd been gifted such an opportunity, but then America improved my opportunity to sample all kinds of things, from varieties of peanuts in the minimarket (17!) to beers on tap (67!). As a counterpoint to such freedom of choice, there were conspicuously more rules. There are beaches in Los Angeles where you need to ask a lifeguard's permission before throwing a Frisbee. Threats for disobeying these rules involved fearsome things: fines, solitary confinement, death by steamroller. My well-being was no longer just my problem but everybody's business, and America demanded that I feel endangered at all times. Example: a signpost ... 'You are now entering a Tsunami Risk Zone' (I'd been blithely riding the coast of Central America for months, a vast Tsunami Risk Zone – what was I thinking?). During my brief stay in Los Angeles, I was warned by staff in a planetarium in the Hollywood hills about the threat of motion sickness (presumably someone had once felt mildly nauseated and had sued). It should be noted that warnings like these exist in a country where you can purchase a bacon doughnut, and can do so without health insurance.

Behind Santa Barbara, I found a small trail called Camino Cielo ('the sky road') that wound up into the Santa Ynez Mountains, and I was soon struck by a lovely, collecting wildness. In the afternoon I rested to eat an apple outside some caves where 400-year-old paintings had been crafted with ochre, charcoal and powdered shells by the Chumash Indians who lived in these hills long before the freakish crowd of modern-day California moved in. Visitors had signed the guest book, and one entry read: 'We're on a bachelor party! Caves were great! Now beer and titties!' The entry ended with a sketch of a woman with breasts large enough to impede standing up, which highlighted, as well as the Indian cave paintings did, mankind's propensity to explain through art. Unnecessarily, perhaps.

When the sun reached its full height the smooth bark of the manzanita gleamed, but then a hot wind swept up a gauze of dust and smudged everything to pastel. Soon the trail was riding a ridge, just a few metres across. The ocean spread out to my left, and to my right a drop-off of tangled scrub met Cachuma Lake, its waters a weaker blue. I wished I could ride a trail like this all the way to Alaska, away from the urban racket and – it often felt, in a cranky Thoreauvian way – all the teeming mess we'd made of the world.

The road turned to dirt the next day and the air smelt lightly herbal. A hummingbird, wings ablur, jinked past my handlebars. Two crested caracaras on a bough calmly watched me crash off the trail and through a bush. When I'd stopped shrieking, I noticed that the sounds of humanity had dropped away entirely, leaving just the ambient trill of streaming bees. The thought appealed to me as – pew, pew, pew – three shots from a gun club shattered the drowsy flavour of the day.

Highway One felt overrated now but the *camino* didn't stretch to Seattle, so I dropped to the coast again and moved along with RVs, all branded with names in stark contrast to the creature comforts within. The Expedition and The Adventurer contained fridges and smoothie-makers. There was the more tepid Excursion, which invites the question – why do you need a 33-foot mega-vehicle with leather couches and a Sony home cinema system if it's only an excursion? There were predators on the prowl too: The Puma, The Cougar, the latter with a bumper sticker that caused me to face-palm: 'USA: Back to back World War Champions'. Altogether less ostentatious was The Mallard, and from behind its windscreen a bespectacled couple, possibly ornithologists, peered out vigilantly. Later that afternoon I was almost swiped by The Intruder, which I imagined sold on the back of an infomercial that began 'You wanna crush some nature? You wanna kick the shit out of the wilderness?'

*

Soon, Big Sur: that famously feral hunk of coast, thrashed by Pacific breakers: a strong choice for a guidebook cover. The road clung desperately to California now, wrapped around cliffs, held up by bridges. Nearly inaccessible once, Big Sur was one of the last places to be settled in the States, and for the pioneers, a trip to Monterey or Salinas took three hard days by horseback. As late as the 1920s, only a few homes in Big Sur had electricity. Henry Miller famously lived here, to escape 'the air-conditioned nightmare', but that was all a long time ago, as was Hunter S. Thompson's assertion that Big Sur had become a rapidly commercialised playground, overrun by bored urban socialites complaining that it was 'nothing but a damn wilderness'. *A damn wilderness*: sounded all right to me.

The old ranger's hut looked abandoned and good cover for my campsite – doubtless the same conclusion drawn by the biker I found behind it, laying out his own tent. With his yellow Lakers shirt, local accent and professed Berkeley roots, I'd assumed that Nate was out for a day or two, max. But he'd been on the road for two years, cycling mostly through Asia, and now that he was almost home, I sensed he was feeling something like grief. Perhaps it was his reflective mood that made that evening one for swapping stories.

The following morning, Nate wandered into the bush behind me and returned holding two slimy yellow things between his fingers, one in each hand. 'Banana slugs!' he said, handing me one. I took it.

'Go on, lick the slug,' goaded Nate.

'Nate, I'm not going to lick a slug.'

'Come on, man, lick it. You have to.'

'I don't have to.'

'Just a quick lick.'

'Will I get high or something?' I imagined the hallucinogenic toads of Mexico, concerned that this would make cycling a busy, vertiginous highway a somewhat lethal experiment.

'No, no, no. Just lick it. Look, man, if you don't feel completely welcome in California yet, it's because you haven't licked a banana slug.'

'I feel welcome, Nate.'

He looked crestfallen.

'Oh for Christ's sake.'

I licked my slug. Nate licked his slug. He grinned like a four-year-old.

'Welcome to California, dude!' He turned serious. 'Can't believe you just did that.'

We rode together until Monterey where Nate branched off. A couple of days later I found San Francisco in fine festival mood. It was staging a foot-race-come-party, 'Bay to Breakers', and from what I could tell, it didn't encourage clothes. Or maybe that was just San Franciscans. Nudists had long congregated in the city, and there were plenty of nude beaches: people stripped off in plazas or even rode the bus or train in the buff. Until recently, it had been common to see naked men carrying only a small towel, not for covering up, but as something to sit on when using public benches, as was the custom. Beacon of liberal tolerance though it is, San Fran's city officials finally grew fed up with the mounting numbers of nudists, and in 2012 a bill, narrowly passed, made it illegal to show your 'genitals, perineum or anal region in public'. Your perineum! Say what you will about San Fran, but I find it strangely reassuring that there's a city with enough activists to fight for the right to expose perinea.

Certain festivals, such as Bay to Breakers, were exempt from this law however, and as I was sitting alone on a patch of grass, men and women, but mainly men, wandered unclothed down the street, thankfully none of them flaunting their perineum too avidly.

A building across the road, several storeys tall, caught my eye next. I'd noticed a group of people partying on its roof and I watched them idly.

And then something fell.

Something shaped like a person.

A mannequin, perhaps. This may sound like an odd assumption, but launching a mannequin off a five-storey building to freak out the people below seemed credible, as almost anything can at a playful gathering in San Francisco. There were a few pedestrians close to where the figure had landed. They did nothing at first; like me, they stalled, watching the pavement where something now lay quite still, something like a mannequin, all of us waiting for the punchline. Then mayhem.

I took to my feet. I was around a hundred metres away, and yet there was already another doctor and a paramedic beside the body of the fallen man when I arrived. He was young, unconscious. His breathing was laboured. As the paramedic stabilised his upper spine with his hands, I glanced up at the roof, and felt dizzy and sick.

It took a couple of minutes for paramedics to siren in. A

group of the man's friends had run down from the rooftop and now a knot of them huddled ten metres away, looking over, not daring to get nearer and discover that their friend's injuries were too severe to survive. They called out his name. In the background, a girl began to wail.

The next day, I picked up a newspaper and learnt that he'd died. I'd seen other, cruelly abrupt deaths in my time in the emergency department and I presumed the memory of this man would blend into the rest, but this was not the trauma call of the resuscitation room. It unfastened something in me.

In the emergency department, it's often expedient to blot out thoughts of a patient's intimate world and loved ones, to conquer emotion and get the job done (for self-preservation too). Doctors and nurses are experts at the art of distancing. Occasionally something can rip through the thick skin of a whole department, and while it's rare to hear a voice over the tannoy announce 'paediatric cardiac arrest', an echoing silence can fall when it does.

I recalled a patient I'd seen one Sunday afternoon in London, an elderly lady, in her eighties, who'd become unwell in church. It started with a headache, which was sudden, severe and unusual for her. Now she was dizzy and vomiting, and my spider-sense was up, the little shiver of foreboding that nurses and doctors all know very well. I noticed she was a little drowsy too, and when I performed a neurological examination, testing her cranial and peripheral nerves, she was clumsy and unable to touch my finger with hers when I held it in front of her face. She had good strength in her limbs, no facial droop and her speech appeared normal, but she'd lost a little of her visual field. I rushed her round for a quick CT scan and waited in the viewing room to see the images in real time. On the screen, white areas appeared in the basal ganglia. The haemorrhaging was widespread. I took her to the resus room and her conscious level dropped over the next half an hour; soon she was unable to find the right words, and then unable to respond to voice or pain. After liaising with neurosurgeons, a decision was made to palliate.

At the end of my shift, I had to 'hand over' my patients to the incoming doctors. I summarised who was in which cubicle, branding my patients by their diagnoses. Bed four was an inferior myocardial infarction, bed five was a haemorrhagic stroke. She'd lost her identity already. I tried not to think about the conversation two hours ago when she was awake and able to speak. It didn't help me or her to linger on the game of bridge she would no longer play on Tuesday, or who might be losing a friend. She died, and she didn't stick in my mind particularly, until now, writing this. The next shift, a man with a ruptured aortic aneurysm died too.

The resus room in particular, with its microclimate of disaster, feels divorced from the world at large. I often urge myself to remember that for many people, the day they come to the emergency department is a vital one, sometimes life-changing. It is rarely life-changing for the staff on shift. It can become instinctive for medics to neglect the personal details of their patients' lives, and the rippling agony of their illnesses (and, in some cases, their death). Shutting out humanity is far harder when relatives are in turmoil, right there, shell-shocked in front of your eyes. It catches me every time.

I can't recall the fallen man in much detail now, but I see his friends with a clarity that's harrowing. They are changeless and horrible in their shock.

*

North of San Fran lies Marin County, a leafy, moneyed place with streets named after trees, where middle-aged women wear flowery dresses and walk pampered-looking red setters and Afghan hounds. Grand houses are half-hidden behind rhododendron, and I passed cafés selling sandwiches at a price that ensured I'd stare longingly at them through the windows, like some Dickensian street urchin.

Above Marin, engineers decided against running State Route 1 along the coast. The land here is densely forested and abrupt. A shoreline highway would have meant negotiating the King Range, so the road was steered inland instead and 'the lost coast' has stayed wilder and less developed. Riding the maze of redundant logging roads, despite their formidable grades, was a better life than the highway and I felt at peace for the first time since Camino Cielo.

It lasted until the small town of Petrolia, where serious police officers were riding quad bikes, or pacing around with dogs. Three weeks before, two hundred miles away inland, a mother and her two children had been found dead from gunshot wounds in their home. Shane Miller, husband, father, survivalist, ex-convict and prime suspect, was missing. He'd grown up around these woods and the police had located his car two weeks ago, a few miles away.

I camped in the woods that night, and, in case Shane Miller found my tent, I slept with a penknife beside me, though the blade was pathetic, I noticed, only good for evacuating belly-button fluff. Stabbing? Not so much. It would be a vigorous tickle at most. Ten minutes into my evening, a siren blast sounded. I had a saucepan of water on the boil when everything flashed police-light blue. A voice from outside: 'Get out of the tent!'

I stepped over my saucepan, into the blue-flickering night.

'Sir, what's in your hand! Sir!'

'Spaghetti!'

'Put down the spaghetti!'

Slowly, I bent over and laid the packet of spaghetti on the ground. There were two police officers staring me down, feet wide apart, hands hesitating on guns in holsters.

'What's that!' One of them yelled, pointing at my feet. I looked down. 'Some broccoli. I was going to make a cheese sauce ...'

'Sir, step over here where I can see you. Sir, now!'

I began to move to my left.

'Hands!' they screamed in unison.

'Okay, okay! Fuck!'

I stammered, something about the dark, lost, tent, tired; I mentioned cheese sauce again. The officer scrutinised my passport and then my bicycle, turning sympathetic at last. I was told to sleep well and 'vacate before sunrise', but in a tone which implied that a SWAT team would be mobilised if I made a cup of tea first.

*

I was soon back camping in state parks where I was occasionally adopted by the occupants of those RVs, cheery families who toasted marshmallows for me, even – actually, *especially* – the ones in The Intruder. ('You never had a s'more! Get him a s'more!') The camping was cheap, though Salt Point State Park was not cheap enough, it seemed, for Adam.

'Hey, you wanna camp with me and my girlfriend? Be cheaper if we team up, they charge per party.'

I sat with Adam and Kiley around a campfire. They'd left their home in Colorado a few weeks ago for the promised land of California, with their dog and a few belongings stuffed into a rusty pickup. They had no work lined up, no friends here, no house ('yet'), and were hoping a flat would come up on Craig's List, but they were broke and Adam's hustling had made them little so far, partly, I assumed, because he was misjudging his market – several times already he'd tried to sell me car parts and wine. I overheard him on the phone to a friend: 'No way, man, this is California. I can't be like selling meat from the back of my van like I did back home. Seriously, man. People want *menus* and shit.'

When Adam was nineteen, he'd been sentenced to five years in jail after being caught with 45 pounds of weed and 8 pounds of cocaine. He'd been transporting it for his father and at trial had decided against ratting out his relatives. He could get work like that again, he told me, $40,000 per job, but the stakes were higher now, they'd throw the book at him next time. He sounded tempted though, especially after yesterday: he'd left his ID and most of their money in a toilet cubicle. Adam told me all this without concern, as if he expected the universe to wink and he'd be blessed. I'd felt the same way myself at times, more so as I pedalled through the world, but especially here in California, which breeds optimism: blame the gold rush, Hollywood, too much sunshine.

Adam picked up a jerrycan, stood over the campfire and tipped petrol into the flames, which shot up into the container, setting it alight. He swung the blazing can around to extinguish it, with Kiley shouting 'Adam, shit! Fucking shit, Adam! Ha ha! Motherfucker!' and then leaping up, heaping sand onto the jerry, putting it out. She was in hysterics. 'Man, that was badass!' An hour later, Adam sat frowning at the last few flickering flames of the fire. He reached forward for the jerry and tipped out more petrol. Up went the jerry again, and up shot Kiley, her fists full of sand. Maybe Adam and Kiley would be just fine.

The hills rolled on, long and sweeping gradients into Oregon. In need of a rest, I cut across to Portland, one of the best American cities to rest in, apparently. I knew this because

I'd watched *Portlandia*, a satirical TV show that invoked a city of people who talked about getting tribal tattoos and forming bands, and who worked a couple of hours a week at a coffee shop. Portland, people said, was a city where young people went to retire.

At the time, Portland was gearing up for a citywide festival of all things bikes called Pedalpalooza and its headline event was the annual naked bike ride. Naked rides take place all over the world, but most are fringe, daytime events, dominated by old hippies and pudgy naturists. Portland's was different: it was probably the biggest on earth for a start, it took place after sunset, drew plenty of young, beautiful people, and almost everyone in the city knew people involved, even if they weren't prepared to strip themselves.

I'd been invited along by Becky, a friend of a friend. We sat naked on our bikes in a thick crowd of naked cyclists. Apparently, there were a number of sanctioned reasons to ride naked around a metropolitan centre through a corridor of fully clothed spectators who were filming us on their phones. Something about a fossil fuel protest, or rights for cyclists, or our right to bear flesh in a non-sexual way. Maybe I was protesting clothes, I don't really know, but before I could decide, Becky saddled up and pedalled off, suddenly all bum.

I rode into the space she made. Electronica boomed from trailer speakers, and naked hipsters streamed around us. By an overpass, people were dancing naked around sound systems so we ditched our bikes and leapt into a naked rave, flapping around until the sun came up.

I got no rest in Portland at all.

*

It took a few days to reach the Canadian border after putting my kit back on, but where the hell was my cheerful Mountie with a maple-syrup moustache? Instead, the border guard was a grumpy Scottish skinhead who was wreaking revenge on travelling society, one cycle tourer at a time.

'Here's what's happening ...' he began, which is one of the last things you want to hear said under any circumstances, ranking right up there with 'how honest do you want me to be?', the manner in which I was once dumped by a girlfriend.

'*You*,'

He pointed at me.

'... have to prove to *me* ...'

He pointed at himself, in case I had trouble with pronouns.

' ... that we're not going to find you working in a bar. I want bank statements, I want papers, I want evidence. Show me what you got.'

Maybe his mistrust was piqued by my passport photo, which suggested someone with several restraining orders. My beard is patchy. I'm sneering. I look as though any minute I'll scratch my chin with a piratical hook-for-a-hand and get back to sex offending. I wouldn't give me a library book.

I had no documents to prove financial security – though my issue was less the lack of documents than a total lack of financial security. I'd not yet met Mary-Ann, my mum's cousin living in Vancouver, but when I phoned her up, she had the grace to take full financial responsibility for me. I was now allowed into the border guard's adopted country, and I'm sure Canada was proud as hell that he was protecting the nation with such Scottish grit.

Despite this introduction, I was fairly sure that I'd have a wonderful time among the amiable Canadians, at least until I was ripped to shreds by a bear. There seemed to be two extreme schools of thought in Canada in regard to bears and my prospects of being savaged to a wailing death. Some likened them to big, curious dogs – hungry and snuffling sure, but hardly worth panicking about, just don't leave food in your tent at night. But there were alarmists too, morbidly fascinated by a grizzly's ability to outrun you, outclimb you, outfight you, chew heartily away on your bone marrow and then maul the toothpaste you left in your tent.

Following the advice of city-dwelling Canadians in the latter group, I shuffled into a small hunting store in downtown Vancouver. My plan was imprecise but had something to do with bear deterrents, whatever those were. The man behind the counter wore a name tag on a camouflaged jacket that said Jake. Serious nose, serious beard. I imagined that Jake had taxidermied his own grandparents and mounted them to the wall of his cabin. He'd have finished them off with a crossbow when they were too frail to haul wood.

'What do you need, friend?' he said, and returned to chewing something.

'Something for bears.'

'For bears?'

'Yeah, for bears.'

He sniffed. 'You'll want some of this,' he said, picking up a can of bear spray, which is a bumper can of pepper spray, the size of a can of paint. I'd already considered arming myself with bear spray, but wasn't sure I could use it successfully under the pressure of a bear attack. There seemed to me only one thing more disagreeable than getting mauled to death by a bear, and that was accidentally spraying myself in the face with extra potent pepper spray, and then getting mauled to death by a bear.

'Twenty-seven-foot range on this baby,' continued Jake. 'Just blast the grizzly right in the snout, kay?'

'Right. The snout,' I said, wondering how close I would have to be to a grizzly bear's snout before I swooned from fear.

'We got these bear bombs too. You want some?'

'Bombs?' I said, in a faraway voice.

Jake looked at me as if I was being ravaged by a wild animal.

'Oh, and we got your bear guns. And your bear bells ...'

'Bells?' I wondered: could I frighten a bear away with a Morris dance?

Jake was busy lining up all these things on the counter so I snatched up the bear spray, leaving behind the flares, guns, bells, bombs and projectiles. Jake bagged it for me.

'You know how to tell the difference between black-bear and grizzly-bear shit, right?'

Obviously, I didn't.

'No? Black-bear shit fulla berries and squirrel fur. Grizzly-bear shit fulla bells and smells like pepper.'

I left Vancouver still very much afraid of bears, but more afraid that Canada would be full of people like Jake. Fortunately, the landscape was calming. Sat amid the ruffled grey-green water of the Georgia Strait, the humps of the Gulf Islands were crowded in pine, and cloud raked at their tops. Towards Lillooet, the land began to look hard and thirsty. They say Lillooet is the hottest place in Canada, a fact expounded by its residents, as if you ought to be stupefied by it. In the queue at a supermarket, a man told me a cautionary tale, but I was only half listening. I think it involved another foreigner on a bike who'd sweated so hard that he was converted into a white crust and had to be scraped off the asphalt before his salt crystals were repatriated. When I pedalled away from the store, it was past a sign warning me not to pass snow ploughs on the right.

A signpost: 'Super, natural British Columbia', a narking slogan that held a sinister truth. Since the seventies more than twenty women had vanished off the main road that sweeps east–west across British Columbia connecting the port of

Prince Rupert and the town of Prince George, most of them hitch-hikers, most of them Native Americans. Signs declared it 'The Highway Of Tears' and the missing women smiled from posters, between tracts of thick, black forest.

From Highway 16 I switched to the Cassiar Highway and Canada got wilder. Fireweed and saffron wild flowers spiced up the treescape and every now and then I glimpsed a black bear bounding away. In the evenings I camped by lakes, hanging my toothpaste in trees and wallowing through reeds to wash off the day. An American cyclist had been attacked by a wolf near here last month. Undeterred by a blast of bear spray, it had chased him down the road and savaged his tent. Local people muttered about how 'incredibly rare' such an event was in a way that didn't console me at all – the psychopathic wolf was still out there. Did it have a taste for tents now, or Lycra, or human blood?

There were plenty of places named on my map but only because almost anything was worthy of a name out here in the monotony of forest – dry creek beds, long-abandoned towns, and other places worthless to me, like lodges with unaffordable rooms and The Rabid Grizzly Rest Stop, which didn't sound like the kind of place I would get much rest in. Dawson was more auspicious: an actual town famous from the gold-rush days, its history writ large in a boardwalk and a nightly cancan show. There's a bar in which people still drink sourtoe cocktails, which contain a mummified human toe, as a crowd chant 'You can drink it fast, you can drink it slow, but the lips have gotta touch the toe!' That's fun, apparently, in Dawson.

I began to chip away at a 19-km climb now, into the Alaskan tundra. The Top of the World Highway began to level out in the afternoon but I couldn't take my eyes off the mountains, folding away blue-tinged and bleary. Somewhere a wildfire had taken hold, and the smoke mushroomed above the forest like a nuclear bomb. After passing through the Alaskan immigration station at Poker Creek (the sign: 'Population 2'), I arrived at a town called Chicken, which had been named after the ptarmigans, or 'chickens', that were once avidly hunted here.* A sign detailed a miscellany of facts about Chicken. The population was fifteen in the winter, thirty to fifty in the summer. The mail came by plane twice a week. The three-legged dog was called Tucker; he'd lost a fight with a GMC truck. He was half collie, half husky and 'he has always lived dangerously'.

There was more of that deadpan humour that I soon accepted as characteristically Alaskan. You could buy bumper stickers ('I got laid, in Chicken, Alaska') and rubber chickens from the shop. Every year the hamlet held a music festival called Chickenstock.

A few miles beyond Chicken, the eastern sky had turned a sinister orange. The Moon Lake wildfire had been burning since a lightning strike in June and the air reeked, not of burnt wood, but of something earthier, more cloying – burning tundra. Mostly, Alaska is left to burn. Around 3,000 square miles go up every year, and some crown fires burn so intensely that they go through the winter months too and are only fully extinguished in the spring when firefighters dig up smouldering earth.

Regardless of when you visit, it's impossible not to dwell on the trials of winter in Alaska. Trials, though, are perennial, and for me, one in particular played out nightly. From an hour before dusk I was imprisoned inside my tent by a fog of chunky mosquitoes. In countless pools of decomposing vegetation on the muskeg, mosquitoes breed and collect in huge clouds. A lather of DEET was only a partial deterrent and I was blotchy

---

* Ptarmigan itself is an odd word and comes from the Gaelic *tàrmaich*, to gather, to settle.

and itching by the time I arrived into Fairbanks, the largest city of Alaska's interior, and a place where, I'd heard, kids in junior school had to play outside unless it was below minus 20 degrees Celsius. Presumably, everyone who grew up in Fairbanks was hard as nails.

I'd need to be too, I thought, embarking on the five-hundred-mile stretch of road between Fairbanks and the Arctic Ocean, the Dalton Highway, or, more colloquially, the Haul Road. It's a supply route for the trans-Alaskan oil pipeline and oil fields of the north slope, but you may know it from Series 3 and 4 of the reality TV show *Ice Road Truckers*, and perhaps you can even recall the show's sensational tagline: 'In the Dark Heart of Alaska, there's a road where hell has frozen over.'

The Haul Road wasn't frozen, not yet anyway, but it was hellish. A mess of grit and mud with the consistency of toothpaste, churned up by oil trucks and gangs of bow hunters in pickups. Sections of the highway had been nicknamed by unfortunate events. I passed Oil Spill Hill, The Beaver Slide, The Rollercoaster and my favourite – Oh Shit Corner, a place where every trucker has had an Oh Shit moment, one told me later. 'Your brakes go out here in the winter and you're at the helm of an 18-wheel toboggan.' The only place with hot food was a truck stop, 265 miles from Fairbanks, called Coldfoot (singular, the other presumably amputated), where burly, bearded men crowded their plates with fried food. I sat gratefully for hours among their denim and heart disease.

Beyond Coldfoot, the trees got spindlier until the Farthest North Spruce Tree (more symbolic than actual, it was marked by a signpost and a plea not to chop it down). From here naked tundra reached northward, sap froze, and permafrost prevented roots from anchoring trees. I climbed over the Atigun Pass, crossing the Brooks Range and the Continental Divide, the mountains all shale and umber grass, their peaks lost in a cold and roiling fog. A headwind rushed hard and bitter and powdered my panniers in snow. More snow fell through the night and I woke to find that the mud stuck to my bike yesterday had frozen, locking my chain, brakes and even my wheels – a particularly galling turn of events in the effing summertime. But even then, with the land ahead of me running out, my luck felt limitless. In half an hour, a truck full of road-workers appeared, angels with power hoses. 'How 'bout we rinse your bike?'

Bike gleaming, I rode off, my shadow drawn long across the tundra, counting down the last miles of the Haul Road. I was inside the Arctic Circle now, where there would be no sun to raise shadows for months each winter. The same shadow cyclist had stretched out mornings and evenings for the last 21 months – I'd watched him darken Mexican cacti and quiver in the Festuca grass of the pampas and he felt like a friend. He embodied the simplicity I was so fond of, the daily cycle of eating, riding, thinking, reading, sleeping – an intoxicating minimalism that exploded now and then with something awesome and unforeseen, like the dark blots I could see now on the horizon: roaming musk ox, wiped out by the 1920s in Alaska and reintroduced from Greenland to help prevent their extinction. Up above, snow geese honked in formation and for a moment I was transported back to the wilds of Patagonia, altogether different in tone and smell and light, but similar in their openness and violent sense of privacy.

Finally, the town of Deadhorse, Prudhoe Bay. I'd assumed that the eponymous horse had died of cold, but after nosing around Deadhorse I couldn't rule out boredom. The town is a base for oil workers, not travellers or citizens, and the best thing about going to Deadhorse is being able to say that you've been there. One quirk of the town is that nobody locks their vehicles, not simply for lack of theft but because there is one exciting thing that can occur here: amid thick fog, grizzlies, and occasionally even polar bears, pad into town. Leaping into a vehicle, any vehicle, is the best escape.

There was a deserted cabin on the edge of town, listing into the permafrost and trashed inside. I cleared away some shattered glass and made a camp in the back room. For the next couple of days I ate for free in the all-you-can-eat buffet in the canteen. Turns out that grime, a knotty beard and a thousand-yard stare are common to both long-distance cyclists and oil workers, so I exploited the mistaken identity.

Fed and rested, I rolled my bike back onto the Haul Road, hoping to hitch a ride on a truck back to Anchorage and then fly to Sydney. A young man in overalls and a beanie wandered over.

'You cycled up from Argentina then?'

'Yep.'

'I see a few of you guys. Why you wanna do that?'

Why?

To lie sleeplessly on salt flats. To be endlessly stupefied by the kindness of strangers. To eat well, or at least copiously, and have earned it. To find space to think, and to laugh to myself about things too small to notice were it not for slow miles. To squat in a ruined Portakabin, very alone, pretty cold, and still feel overcome by the happiness wrought from an adventure.

How to condense all that?

'I wanted to see Deadhorse.'

He smiled, glanced around and shrugged.

'Well, helluva journey. Hope it was worth it.'

Part Three

MELBOURNE TO MUMBAI

As to hunger, a man can live on his own fat for a week
and it is a poor country where there are no lichens.
The Happy Traveller: A Book for Poor Men,
Rev Frank Tatchell, 1923

6

Fever

A bicycle allows for a softer transitioning than most means of going places. Not always – you can still cross thresholds, burst over the top of a valley, or arrive with a thump to somewhere that feels entirely new, but mostly a cyclist doesn't land *in* places as much as they are surrounded *by* them. And as the world changes shape around you, there's a sense of connectedness: time to dwell on the grey zones between hills and mountains, mountains and valleys, valleys and floodplains, nations, peoples.

It can be exciting to jump on a flight and *arrive,* but it's discombobulating if you've been crawling through the Americas for more than 600 days. Heading to Australia involved some stupendous jumps: from the northern to the southern hemisphere, from GMT –9 to GMT +11, from the autumn of Alaska, which can sharpen your nipples, to the Australian spring, which can soak your pits. Perhaps most disorientating of all: I was no longer riding alone. Three years a singleton, I was now cycling out of Sydney beside my girlfriend, Claire. Unlikely, but incontrovertible: there she was, right beside me, riding a shiny-black touring bike and winking at me every now and then. I could even reach out and touch her.

'What are you doing?!'

'What? Oh, sorry. Nothing.'

I was a student when I first met Claire in a club called The

Magnet in Liverpool, a basement den of red vinyl seats, rolling basslines and a stream of celebrated DJs and touring bands. Claire managed the bar and played saxophone, and I spun funk, soul and hip-hop records to sweaty hedonists whose main aim in life, like mine then, was to seize the weekend. We began going out on a Saturday night – by the following Thursday she'd dumped me three times. Afterwards, I'd assumed I was dumped for good, but if a bicycle expands the world, Facebook shrinks it. While riding through Canada, fortified by eight thousand intervening miles, I messaged Claire to ask if she fancied riding with me – across Australia for starters, and beyond, if she felt like it. And then, poof! She appeared in Sydney a few months later and I was promoted to boyfriend again.

We left the city by ferry and headed to Manly, a beachside suburb, a little further up the New South Wales coast. Standing on the deck, I held Claire's waist and let my chin rest on her shoulder, the swooping fabric of the opera house reeling past. A band of men were clutching tins of beer on the quayside. No doubt in response to the love-struck gleam in my eyes, one of them bellowed 'Go on, mate!' and dropped his jeans. A magical moment, almost filmic, was swiftly ruined by a strange Australian penis, flapping about in the harbour breeze.

It was my plan to strike north, riding Australia's eastern seaboard to Cairns. Claire went along with it. Obviously, our love would blossom on the open road and I imagined that we'd be helplessly locked together now, from sunrise to moonlight, all soft words and white-hot passion. It's possible that I was getting ahead of myself. First things first: we were heading into the tropical heat of Queensland, at the muggiest time of year.

Apart from the Australians themselves, little about that summer was kind to us, and the raging, midday heat was the least of it. Famously, Australian wildlife is ludicrously dangerous, having been left alone to play a lethal game of one-upmanship for millennia. For the first few hundred miles north of Sydney, we were swooped by magpies when we got too close to their nests. The naming of the Australian magpie derives from its black-and-white colouring alone – they are not related to the inquisitive but passive creatures I was familiar with from home. Australian magpies are brawny, fat-billed psychos, cousins of the butcher-bird. They whooshed past our ears, snapping their beaks and even body slamming onto our helmets.

A week later, Claire came charging back to the tent. 'Little fucker!' She screwed up her face and examined her left foot. 'I've been bitten. There was a spider in the toilet.'

We were close to Brisbane, camped in a lay-by. Bearing in mind that the offending spider was Australian, I presumed that we could skip the first aid and go straight to palliative care. On the off-chance that Claire wasn't about to turn into a purple ex-girlfriend, we trapped the culprit under a cup for the benefit of the coroner, or the paramedics, if they arrived in time.

The paramedic picked up the cup, casually, sniggered and gathered the spider in his bare hands. It was several times larger than anything I'd seen in Britain.

'It's only a bloody huntsman,' he said, as if the local kindergarten had more lethal species.

'No need to hurt the little fella.'

He walked outside the toilet block and released the spider into the grass, where it scampered off in the direction of our open tent. There was nothing mean or prankish in this. To the Australian mind, a large arachnid plunging its fangs into your naked flesh is no particular cause for concern; these things happened from time to time. He probably had a childhood mate called Robbo who'd lost three limbs and an eye to a rabid galah. We thanked the paramedics, returned to our tent and held each other tight in the dimness. For killing the romance, there's nothing like a sense of impending doom. In the morning, I peeked out to check that all was clear, fearfully scoping the grass for the common death adder: three words I'd been dismayed to find exist in sequence.

North of Brisbane, the air grew thick with flying ants and the sky was stacked with thunderheads. Rain fell with the gusto of a power shower. We dropped our heads, as if diminishing ourselves by a few inches would save us from the next lightning strike. It had been a destructive few years for Australia and it was impossible to ignore the debris of visiting natural disasters – arboreal wreckage from cyclones, scorched earth from epic forest fires, relics of homes flooded near Gayndah. We plunged on, soaked in sweat by lunch and dog-tired by dinner. Claire seemed to be losing the dimples that appeared when she smiled, and 90 per cent of the time there was a sense of being slowly hunted.

Luckily, it wasn't all grim. There was that magical 10 per cent: the backcountry trails. We joked and shared stories as we rode. Living rough suited Claire. Like me, she thought nothing of washing in rivers, and our happiest moments together were leaping nakedly around water holes, like Adam and Eve with tan lines and chain oil on our legs. There was some relief off the road too when we stayed with strangers discovered online through a travel networking website for cyclists, though our hosts would often wonder aloud why we were heading north at this time of year, to which I'd reply 'I think it has its upsides', fall silent, avoid eye contact with Claire, and scratch the March fly bites on my arms.

*

'Jauh' is the Bahasa word for 'far' and I liked its ring, which sounded true to its meaning. I could recognise 'far' in a dozen languages now – *mbali*, *lejos*, *irak* – it was where I'd come from, where I'd been, where I was going. Far had been the world's collective vocal tick since I got past Albania, and then far was far: to a lot of people, it made no difference if I'd cycled 3,000 or 40,000 miles, it was simply a long way to come. Claire and I were almost as far from home as we could be and I felt dizzied by the bygone miles, like a big wall climber, glancing into the abyss.

I felt dizzy too when I considered the task of biking across Asia – there were some meaty challenges ahead. Asia had the most foreign of all languages for an English speaker, outrageous mountains, and countries that were, in fits, turbulent. Several of the visas we were hoping for could easily be denied to us, probably by a dead-faced official speaking through the metal hatch of a backstreet consulate.

After Christmas with friends in Cairns, we flew to the island of Timor and began riding Asia's scattered edge, connecting the islands by boat. I'd given very little thought to Indonesia before and if, like me, you imagine a palmy, trivial place, the stats are excruciating: 13,466 islands, 719 languages, 260 million heads. Jakarta was the most Instagrammed city on earth at the time, and yet the nation hardly ever made international news.

Perhaps its relative obscurity fed the sinister preconceptions about Indonesians themselves. A few weeks ago, back in Australia, the news channels had been covering a story about a pretty, blonde school-leaver (a 'schoolie', in Aussie parlance) who'd become unwell in Bali after her drink had been spiked. She was returning to Australia for ongoing medical treatment. The press were feverish, though of course there's much less interest when drinks get spiked in large Australian cities (by men with lighter skin). Indonesia was the bogeyman next door – look out for your daughters – a mysterious rabble of islands, heaving with potential stowaways, illegal fishermen and Islamic extremists – at least that's how the Aussie tabloids preferred them.

Claire and I arrived into Bali by boat from the neighbouring island of Lombok and began climbing gently away from the bay of Padangbai. The ground rose fresh and terraced on our right and the sun slipped through the clouds creating little spotlights that searched the paddies, switching them from dull to brilliant green and back again. The heart of Bali is ruled by mountains, and the roads on my map trickled down from them, twitching towards the coast in imitation of rivers.

We soon arrived into Ubud, a noisy squall of peace-seeking travellers. Plenty of hawkers, too.

'Taxi, mister?'

The driver lent out of the window and drove his car next to me, keeping pace.

'No, thank you.'

'Why not taxi?'

He looked hurt.

'Because I'm riding my bicycle.'

'But I give you cheap price!'

'I don't need one.'

'How much you think?'

'I really don't care.'

'For you, sir, extra cheap.'

Most of his upper body was now projected through the car's window and he drove on, one-handed.

'Hotel?'

'No!'

'Mango?'

'No!'

'Okay, okay. I know …'

He reached into a bag on the passenger seat, and removed some metallic apparatus, which he held out of the window and jangled.

'Wind chime?'

'Please ... where I am going to put a wind chime?'

He looked defeated, but only for a moment. His eyes popped open again and he dropped his voice.

'Marijuana?'

*

We fell in with the Ubud crowd, browsing the shops slack-jawed, tittering at the wooden phalluses for sale, visiting the Sacred Monkey Forest, where obese, surly apes launched themselves onto the heads of unsuspecting tourists, hoping to steal food, or had monkey-sex (which is doggy style, in case you're weird and have never stared at primates fucking each other) on the temple walls. Just another thing to gawp at in Ubud.

My map hinted at a way out: a wild, isolated slip of a road, a rebel among the trails tumbling to the coast. It twisted through the centre of Bali, from one lake, Danau Batau, to another, Danau Buyan, following the rim of a large volcanic crater.

It was early morning when we left town. The old men attached to the small family-owned cafés, *warungs*, watched us ride out without a flicker, living manifestations of *jam karet* – 'rubber time' in Bahasa – the notion that things will happen when they happen, just wait around and see. Offerings of pressed flowers sat on doorsteps, and the air swam with the dozy scent of incense. A few travellers were up to greet the dawn with sun salutations: white thirty-somethings, floating by in full Bali uniform – sarong, sandals, wispy beards.

Ubud fell away and we began to climb beside a grand-stand of rice, sensationally green. Outside villages, dogs nosed through garbage in packs. I don't fully understand why this is, but from 200 metres a dog can pick out a cyclist from a line-up of pedestrians, cars, other dogs and livestock. It's uncanny. If the chase was on, I stopped and mined my pocket for stones. Generic rules apply – it's 5 points for a body shot, 10 for a head, 20 for a snout, and 50 for inflicting permanent damage. I went for the jackpot every time.

We passed men standing shin-deep in the rice paddies, smiling with great red-stained lips from chewing the betel nut. Children ran at our wheels – a happy, hectoring comet tail – shrieking 'I love you, mister!' and once: 'I never forget you, mister!'

The asphalt began to crumble and the road got steeper, but our vantage point improved and below us hills deepened, parted by snarls of forest. We were heading towards the layered reliefs of mountains now, and mist wafted creepily in fingers around Mount Agung. The volcano was beheaded by white cloud, but it looked peaceful, far removed from the pyroclastic tantrum of 1963 that cost more than a thousand lives. A disturbing thought: if it fired again into life, if the eruption was anywhere near as violent as those of Rinjani and Tambora, Bali would be a dead zone.

We topped a thousand metres above sea level, in cool, cloud-soaked air. Wooden shacks were lost in an overgrowth of papaya trees, spiky red rambutan and mangosteen. By late afternoon the trail was alarmingly steep and in three hours, we'd moved a measly ten kilometres. I watched Claire submit to the mountain, stamping down on the pedals with great, antenatal huffs that rivalled my own. She stopped now and then, crumpled over her handlebars and looked up at the rain clouds in a rosy-faced, religious way.

'I feel like trifle,' she said.

Hard miles pushed me into a sullen silence, but Claire talked in metaphors that veered towards the culinary. So far, she'd felt like 'custard', 'warm yoghurt' and once 'a trout swimming through strawberry jam', but she'd never reached the trifle stage before and I was worried. It sounded like it might be worse than custard.

We found the lake, or, rather, its veil of mist. We began to head down, passing a small concrete house where musicians with thin moustaches were sitting cross-legged on the porch. One was blowing into a *suling*, a bamboo flute, its warble dancing over the rhythm of a kind of bamboo xylophone. The men called us over to sit out of the rain and a shout went up for *arrack*, the local firewater, fermented sap of sugar cane. It arrived with a woman in a pale kebaya and dashing red sarong. I remembered Claire's flute so I began nudging her in the ribs until she said 'Oy! What's with the nudging? Okay, fine, I'll get it.'

The *suling* went silent. Claire's hefty metal instrument was an alien here and she was no expert on the music. The flautist in particular looked narked. But slowly, the men began to play again. Claire listened, eyes shut for a few bars, and then began to play too.

She nailed it. The flautist's melody was resurrected to a note. The other men were glowing with surprise, everyone cheered. A jam got underway, flute to flute, his call, her response. 'Arrack!' they cried, and the afternoon fell fuzzily away.

*

Bali is home to over four million people, and a taxi driver in Denpasar, the island's capital, helped clarify why. 'I moved here from Java years ago. Better money! In Bali 1,000 rupiah is nothing. It's candy. In Java, people are killed for less.'

But people means traffic and we were soon slowed down,

boxed in. They say that to be a safe cyclist you should expect the worst. In Asia, and especially Indonesia, that kind of thinking will get you killed. Reality is far worse than you can imagine; it's supernaturally scary out there. Air horns and blind faith don't make for an effective highway code and the number of road deaths in Indonesia is eye-watering.

A microlet minibus pulled up beside us. Swirled across the chassis: garish colours, an image of James Bond and 'VIP Class'. As we were both held in traffic, I took a peek through the passenger window of the bus, to the inside of the windscreen. It had been covered, almost entirely, with toy animals and suckered-on knick-knacks. The driver was a twitchy adolescent with a red Mohawk and, sharing his view, I watched a motorbike disappear behind a yellow caterpillar. The ambulance up ahead flickered in and out of plastic birds and the tentacles of a purple octopus.

The minibus changed lanes, drew alongside us, and then braked hard to pick up more passengers. The bus was already heaving, but still men, women and children wedged themselves inside, head first into air pockets, until the front seat was a block of intersecting limbs, like a mass grave. Dangdut music blasted from the speakers, jet-plane loud, and the vehicle lurched off again through a battlefield of potholes. The other road users didn't inspire confidence either – half the motorbikers looked like they'd graduated straight from something plastic and Fisher-Price to Suzukis with more horsepower than they'd had bedtime stories. The air ponged of petrol fumes, grilled meat and rank vegetables; a powerful cocktail that made me floaty-headed. At least I'd be pre-anaesthetised for the pile-up. A motorbike driven by four children raced past us – there was a live pig in its sidecar. You can't unsee something like that.

To celebrate our survival, against the odds, we chose to

leave Bali immediately. Java was the next island in the chain and Indonesia's most populous. Claire and I were hustled out of Soekarno–Hatta International Airport in Jakarta and into a waiting car. Simon, a British CEO of an insurance company, a stranger who'd been following my blog, had everything covered for our time in the city, from Javan cuisine to chauffeurs and accommodation with his friends, Anne, Phillip and Zoe.

We'd have to get to them first. Having pedalled through Istanbul, if such a feat can be said to be possible, it was hard to comprehend a city with more gridlock. Jakarta is that miracle. It is one of the world's greatest metropolises but its underground network is disgracefully delayed (by thirty years, according to old feasibility studies), so short journeys could take many hours. It's probably the worst place on the planet for traffic, and in 2017 the jams were said to be costing the city $5 billion a year. To battle this nose-to-tail nightmare, laws were passed to encourage motorists to have a certain number of passengers in each vehicle, but the policymakers hadn't bargained on 'jockeys', men and women who stood by the highways and jumped into passing cars, staying for a small fee, allowing the drivers to avoid the heftier fines. Inventive and entrepreneurial, sure, but you had to feel for the politicians and city planners: plug a hole, the thing keeps leaking.

Anne volunteered in a school built for the children of Jakarta's rubbish pickers, who lived on a bloated, unmanaged dump called Bantar Gebang, in Bekasi, the east of the city. The school was run by Irina, an Indonesian doctor with fine black hair and a wide, lipsticked smile. She invited us to visit the dump and to speak to the students about our ride. The idea appealed to me – it was a chance to reflect on the obstacles to health particular to life on a dump, the physical cost of living on the sidelines. But more than that, I felt compelled to seek connections between an Indonesian slum, a mobile clinic in an African desert, the cubicles of St Thomas' A&E and whatever else lay ahead; some kind of universal truth, perhaps.

We visited the school first, a tidy sanctuary within the dump, with a wooden pavilion, a well-clipped garden and a fringe of elephant grass. Inside, an old man was sweeping dust with a palm frond. Kids hurtled around, until, wide-eyed, they stalled at the sight of us and whispered to each other with cupped hands. For the children who did attend school – who could abide schedules and timetables and weren't sent to beg or sing for handouts in the shopping districts of the city – 'homework' meant sorting litter on the dump.

We talked to the students and showed photos of distant lands, mountaintops, deserts and salt lakes, and though I wondered at times if we were communicating much more than a lesson in the opportunities of money and privilege, they were bursting with questions by the end and I felt glad that we'd come. And then we went walking through their world with Irina, tiptoeing between puddles, through gritty mud, and down paths made of spongy cardboard packaging.

The homes were constructed from plywood, bamboo and metal sheeting, with truck tyres on the roofs to stave off dismembering gusts of wind. I glimpsed women squatting, sorting through the rubbish, an eye clouded by a cataract, the raw macules of a skin disease. Everywhere you looked, stuffed sacks of plastic bottles rose in piles and banks, and beyond it all, the horrid loom of rubbish, a stadium of it, and here a few pickers and ratty dogs grubbed over scrap metal, bicycle parts and damp mattresses green with mould.

Clashing stinks resolved into a potent, sugary reek of decay, trumped, occasionally, by acrid plumes of smoke from burning rubbish. The numbers could stun you as much as the stench: 830,000 people lived within 10 kilometres of the dump and it swelled by 6,500 tonnes of rubbish each day. It was hemmed in by thriving highways on which more fortuitous Jakartans voyaged noisily, day and night. The dump might be seen as a symptom of Jakarta's own bloating. It's one of the world's megacities, a place so populated that it's hard to know precisely where it begins and ends.

Some of the 6,000 *pemulung*, or rubbish pickers, were casualties of natural disasters, especially floods, which washed away homes on villages in Java. Others had once lived on agricultural land that had been sold and given over to manufacturing. Without skills befitting the workforce, poverty began to gnaw at communities, thrusting many towards the big city.

The slum worked on a complicated system of bosses, loans and perpetual debt.

'A lot of people here say debt is good – if they had no debt, they would have no motivation to work!' laughed Irina, and I recalled what Orwell called 'the great redeeming feature of poverty' – its tendency to annihilate the future.

Going rates were 2,000 rupiah for 1 kg of plastic bottles, and 5,000 rupiah for 1 kg of glass bottles. In the nearest medical clinic to the dump, I found a price list too: 40,000 rupiah for a medical check-up; 900,000 to deliver a baby; 1,200,000 if you wanted intravenous medications with it; and more if you fancied a birth certificate.

The resident threats to health could be guessed at: fire and slips of rubbish. The classic scourges of poor sanitation were here too of course – diarrhoea, skin infections, disease carried by insects and parasites. Irina spoke of other corollaries of poverty.

'Some girls become mothers at thirteen or fourteen and often deliver in the slum, boys get circumcised here too. We get infections. Some girls wander around the dump at night and pick up guys, the guys pay, or offer to pick litter for them. One of the girls told me that she charges 5,000 rupiah, about 35 cents. She said to me: 'Cheap, because I have fun too!'

Behind the grime and deprivation, I began to see hints of harmony in life on the dump. A woman sat behind another, plaiting her hair. Three girls trampolined on an old mattress. Small shops were tucked away in the alleys, doing gentle business. Away from the loftiest peaks of junk, a little breeze had kicked up, lifting the lid on the oven, scattering hordes of flies and letting three boys take turns to fly their kite, which flitted against the wavering refuse. I'd taken the dump as a chaos with living spaces mined into it, but I could see now there was order, even ornaments: a pastel-coloured lightshade but, of course, no bulb.

In one shack, I met a woman in an orange kebaya, mother to four children. She didn't know her age. Last year, her four-year-old daughter had come down with a fever. Viral infections like dengue, transmitted by mosquitoes, were rife in the dump. She'd taken her daughter to the local hospital but with no money and no documents, nothing could be done. Her daughter's condition had deteriorated, and after drifting into unconsciousness, she'd died. Her expression was impassive as she spoke, not mournful, so despite the story I was surprised when she wiped a tear from her eye.

There are perhaps 100 million dengue infections globally each year, and thousands of deaths, though some people are more susceptible to severe disease than others, and the statistics deserve some context. 'The microbe is nothing, the terrain is everything', said Louis Pasteur. He was speaking of the body, and the bodies and immune systems of thin, vitamin-deplete children living in rubbish aren't the most resilient.

What we know now as 'tropical diseases', scourges affecting those in a band around the middle of the globe, were not always so. In the seventeenth century, in the southern marshlands of Shakespearean England, there was a disease known as ague, a mysterious plague that some historians claim had a mortality comparable to that of malaria in sub-Saharan Africa today. The disease is mentioned in several of the bard's plays. Ague was probably a conglomerate of infections that caused fever, of which malaria (probably the *vivax* form) was one. An odd thought: that malaria was once endemic in the Fens, the marshes of the Thames Estuary, south-east Kent, and parts of Somerset, Cambridgeshire and Lancashire, and especially ubiquitous in the warmer summers. 'Borough ague' described a fever prevalent in Southwark Marshes near Guy's, the hospital I'd trained in. It was well known at the time that people living in the marshlands were sallow, insalubrious and stunted. Travellers to Essex in the eighteenth century, like Daniel Defoe, were shocked to see whole families quivering with fevers in hovels.

An old Kentish proverb went:

'He that will not live long. Let him dwell at Muston, Teynham or Tong.'

There are several clues as to what ague might have been. 'Ague cake' was an enlargement of the spleen, which sometimes characterises malaria. The fever often responded to 'Peruvian bark', chinchona, which contained quinine. Thomas Sydenham (1624–89) in *Observationes medicae* in 1676 described tertian and quartan fevers and related them to insects: 'When insects do swarm extraordinarily and when ... agues (especially quartans) appear as early as about midsummer, then autumn proves very sickly.'

As rural populations declined in Britain and the marshes dried up; as quinine became more available and as poverty eased – so did the fevers. The last endemic case of malaria in the UK was in the 1950s and Europe was declared free of malaria by the WHO in 1975. Mortality in parts of southern England used to be high for the same reasons that Indonesian slum kids die from dengue. The English living in Kent and Sussex in the seventeenth century were desperately poor, malnourished and living in unsanitary conditions, cheek by jowl. Often, they were co-infected too since TB and other infections were rife. Poverty is a great enabler of disease and they were good terrain.

*

From Jakarta, we flew to Sumatra, the next island in the chain, riding through Padang, a city pinched between the ocean and the faint bulk of the Barisan Mountains. By now, we were both struggling through our relationship. Perhaps we were doomed from the start, though bolts of infatuation don't provide for that kind of clarity. We'd become cohabiting partners, more or less overnight, and spent almost every moment of every day together. I hadn't adapted well. Four years of me-time had probably made me more taciturn and self-dependent than is healthy for any relationship. The stuffy heat of Indonesia felt not unlike what was happening between us. Privately, we wondered if things would work out, and, privately, we wondered if the other was privately wondering if things would work out. All the privacy made for occasional flashpoints, and they were getting ever more banal.

'You always choose the campsite.'

'So you're not going to oil my bike too?'

'Well, if you're gonna drink a litre of Molten Caramel Max milk, you will feel sick.'

I drove us on anyway. Mulishness was how to succeed in a bike ride after all and I treated our relationship the same.

After asking directions to a hotel in Padang, a policeman on a motorbike escorted us at cycling pace for ten kilometres. Such little acts of charity had become commonplace during our time in Indonesia and I'd noticed myself less and less surprised by them. More policemen waved us down as we left town too, keen for a photo. This was not unusual in Indonesia, a snap-happy nation. I've always thought that there's a kind of justice in this. For decades, western tourists have been the assailants, ruthlessly photo-documenting their travels and decorating them with local people. Nowadays, with the global profusion of camera phones, I'd been photographed scores of times a day, from Sudan to Syria, with and without my consent or even awareness. At least these liberties were taken on home soil.

We climbed from Padang to Lake Singkarak, the water cherry-red in the sun's flagging light. Fruit bats flipped about our heads as we shadowed the eastern shore. A small boy had captured two cicadas in a bottle, which makes them go loudly berserk, and he waved us down, eager to show us his naughtiness. It was Friday, and young girls in white chadors were heading to the mosque, clumped together like mushrooms. The posters for the upcoming national election were Islamic green now, the candidates in prayer hats with the haj in the background. That evening, I heard the voice of an imam, a flood of Bahasa, and the words 'Europe!' and then 'America!' and I hoped the context was more positive than I imagined.

As we approached the town of Panyabungan, Claire was making long work of the hill, meandering up it like a drunk. She had a bleak, pasty look. I could see straight away that she was sick. 'I'm sick,' she said, hunched over, and then was, all over the road.

We found a guesthouse, its lobby thick with stale fag smoke, a soap opera on the TV – all overacting, abrupt dialogue, distant stares, and weird supernatural sequences that the actors themselves seemed troubled by. The place was run by a bunch of sweaty men, the owner in a T-shirt depicting Osama Bin Laden in the foreground, with the burning Twin Towers beyond. We were given a room, which was invaded by the whine of traffic and the call of geckos scrambling up lime-green walls (the international colour of crumminess). Claire made for the bed, tumbled onto it and was lost to the sheets.

'Police! Open!' It was past midnight. I stumbled to the door, knuckling sleep from my eyes. Two officers stood sternly outside. 'Search!' one said, peering behind me and pondering the lump of sheets. Claire raised her head.

'Married?' he asked.

'Yep. Sure.'

It occurred to me, and not for the first time, that my chosen guesthouse was doubling as a brothel. Police officers stamped around as we ran through some predictable lines of enquiry: my name, age, premier league football team, religion, feelings about Indonesia, and whether I'd been circumcised. As one officer was peering at a file of papers, he said:

'It's terrible. Do people die like this in your country?'

He flashed me a photo of a real-life crime scene, a mutilated corpse by the side of a road. I couldn't tell if it was a vehicle crash or something less accidental.

'Sometimes,' I managed, and he shook his head in a way that conveyed a profound disappointment in the state of things. The police flocked out of the hotel, taking with them a couple of young women and some very despondent-looking men, including the one wearing a T-shirt commemorating homicidal jihad.

The next morning, half a dozen girls were waiting for us outside, eighteen and nineteen years old, in a spectrum of jilbabs, no two the same shade, as if they'd coordinated. They were students, and someone, perhaps the police, had tipped them off.

'Hello. Can we make conversation please sir?'

We spent an hour in a teahouse being quizzed by the braver and more fluent, asking them questions in turn. They baulked when I asked if any of them were married. 'Not until 25! We love our studies.' Their turn:

'Do you know Kate Middleton? How is she?'

*

Indonesia has a troubling problem: amid some of the most densely populated islands on earth, there are hundreds of active volcanoes, and more here than in any other nation. A colossal disaster, sooner or later, is a certainty. Krakatoa lies just off the Sumatran coast, but its famous explosion in 1883, audible in Perth, is dwarfed by that of Tambora in 1815, on another Indonesian island, Sumbawa, to the south. As the volcanic booming began, armies presumed it cannon fire and prepared for war. Troops were marched from Yogyakarta on Java, and boats were dispatched to find ships in distress. It is the greatest eruption ever recorded in human history and left around 90,000 dead, but it didn't end there. A veil of sulphurous dust enveloped the earth and led to the Year Without Summer. Half a world away, with Europe fractured after a generation of warfare, crops failed leading to famine; there were food riots and a sprawl of refugees. Historians have theorised lavishly on its global influence, pointing to flooding in China, a cholera epidemic in India and typhus in Ireland. Painters of the era were inspired by the tephra in the atmosphere – extravagant sunsets are a feature of works around this time – and in Germany, as crops and horses died, a young man called Karl Drais began researching new ways to transport grain. He came up with a mechanical draisine, or velocipede, the descendent of which I was riding around the world. It's perhaps a little highflown to say that the bicycle was born in the belly of an Indonesian volcano, but I like the notion.

Humans have survived even greater eruptions than Krakatoa and Tambora, however. Toba was a night-foisting super-eruption that took place around 70,000 years ago on Sumatra, and the event is thought to have been responsible for a ten-year global winter. One theory, though not without controversy, suggests that it led to a bottleneck in the human population, perhaps chopping us down to just 10,000 individuals, which could explain why genetic variety is less than expected considering the time we've had on the planet.

We topped a two-thousand-metre pass before rallying down to the site of this Armageddon, dreaming of Toba's explosive history. The land before us plunged into the unflustered waters of the lake, and I quickly formed a notion about the Batak people as happy-go-lucky musicians, for all I'd seen were small groups of men leaning back with guitars, sipping jungle juice, a local home brew. From the uncanny serenity of Toba, it was a simple run for our lives down the highway to Medan, and then a quick hop across the sea to Singapore.

*

The first clue that the Tree In Lodge hostel in Singapore was a kind of sanctuary for cyclists was the front door, which had a bicycle crank arm for a handle. Inside, a scuffed touring bicycle dangled from the ceiling, and in the corridor bikers who'd passed through grinned unconvincingly from a mosaic of photos. They looked like we did: flushed and damp-limbed, squinting to somewhere beyond the camera, a trampled look in their eyes. You can't hide from the heat of Singapore.

The guesthouse was run by Swee Kian, or SK, the sparky, bespectacled doyen of the cycle-touring world who had once pedalled from Finland to Singapore and was a repository of information on the city and on the various routes you might choose to ride across Asia. He whipped us on a tour of the hostel. Tree In Lodge was a compact joint of clustered bunk beds and cycling paraphernalia. Upstairs there was a kit swap: ragged cycling clothes you'd only take if you were desperate. I snatched three pairs of shorts, two hats, a T-shirt and some cycling gloves.

The next morning, I found Claire sitting on a bunk bed, hugging her knees to her chest. She was brooding about something, even I could intuit that.

'Just bought a flight to Japan,' she said.

We'd knocked about the idea of a short break from each other, before one of us rose silently before dawn and cut the other's brake cables.

'It's one-way.'

'But two singles will be more expen—'

And she looked at me, kindly, as I got her gist. For a second, I felt annoyed: we'd been unhappily suppressing our unhappiness together, it was teamwork; just where did she get off, breaking up a doomed relationship before it crash-landed in a blaze of screams and regret? And then some reasonable part of my brain clicked into action and I nodded. It was the right thing to do.

'It is what it is,' she said.

Or more precisely, it was what it was.

We shared a feeling of release after that, and we were completely happy together for those last few days in Singapore. I hugged her goodbye at the airport.

The city seemed to change once Claire had gone; it felt forbidding and muscular now. Singapore is struck by lightning up to 200 times a year and it flared that evening without her, the skyscrapers looming Gotham-esque. The aesthetic matched my mood as I paced the streets, dissecting what had been. I decided that, circumstances aside, it would never have worked out. I don't know if this was a dishonest post-mortem, but I was about to spend a great deal of time alone. It was a long road home and my bike was heavy enough. I didn't need the extra burden of guilt or regret to fill those spare hours, or to wonder if I'd been too aloof or inattentive or headstrong, even if, in my defence, I was out of practice in relationships, and, for the short duration of this one, was covered in sweat, mud and March flies.

I moped along. Supercars stuttered beside me, gold and volcano-orange. Singapore's frequent traffic lights made them feel like another excessive detail of the city. Amid the trappings of wealth, there was an obvious paucity here too: the most towering dilemma of the city is a lack of space. To compensate, Singapore evolves at full throttle. Cranes bow and toil, ceaselessly hoisting new structures higher into the sky, remorselessly scratching out the old so that even the dead get a limited time underground – whole cemeteries have been exhumed to make way for an expanding parade of prosperity. A city of billionaires, 150,000 maids and the highest-paid politician in the world, it's been frisky with the death penalty and liberal with its regulated red-light district; a place where those of Chinese, Malay and Indian descent live cheek by jowl, mostly peacefully, but in a cultural harmony that is legally enforced. It's all-consuming, baffling and heady, which, come to think of it, neatly summed up the last few months of my life too. A city to celebrate, a city to forget.

7
Complications

I left Singapore for Johor Bahru, the final taste bud on the long tongue of Malaysia. Riding north, my view hardly changed for hours. For decades, Malaysia had been the world's greatest producer of palm oil (trumped, only recently, by Indonesia) and reportedly it was now found in as much as half of British supermarket foods. A jumble of fronds stretched thick and distant on both sides of the road. I stared at palm after palm, rows of palms, a trail between palms before more palms gathered and melted into a green blur the colour of palms. I realised that palm rhymes with farm and calm, but lost my train of thought, distracted by the sheer profusion of palms, which were the most palm-like palms I'd ever encountered, they could be nothing else, perfectly palm-shaped and palm-proportioned, exactly typical of palms, and just like palms to be so numerous too.

To make way for the cash crop, Malaysia had cleared vast areas of primary forest. Trucks burdened with timber howled past and there were occasional tracts of barren land, bristled by the nubs of hardwoods. You can feel dismayed at Malaysia, if you like. To do so, overlook the fact that your own country felled most of its natural woodland centuries ago. Then, forget that the spiralling demand for palm oil and timber in countries like yours fuels this deforestation. And best not drag the multinationals into it. They're doing the chopping and they might be from your country too.

That evening, a hot wind blew in, and as clouds massed, nearly black, lightning scribed the sky. There was a hotel with a grubby, broken plastic façade. Still, it looked better than camping in a storm. My room was simple and clean. A sign on the door asked guests not to play with the fire extinguishers, bring explosives into the room, or take the television away as a 'souvenir', and I felt a little sorry for the staff. I amused myself for a while reading adverts in the bathroom for an ambiguous health product called Codi Belle, evidently some kind of miracle cure: 'Meet Nisa. Accident and unable to walk without a stick. After Codi Belle she can walk like normal!'

After locking my bike, I returned to my room to discover a Post-it note stuck to the door. Beneath a doodle of a smiling man standing beside his bicycle were the words: 'Dear Iron Rider, Take care yourself. Enjoy your journey. God bless you.'

*

Jakarta, Medan, Singapore … I was feeling weary of cities and the frenzied highways in between. So now, without research, without deciding on the finer details of my route, I plunged into the rural heart of Malaysia, and on it went: the hypnotic rhythm of palms.

There were little bursts of hospitality here and there. A woman flagged me down and heaped banana fritters into my arms. I stopped for tea and tried to settle my bill but some cunning Malaysian had paid and vanished without allowing me to be British about it.* At food stalls, I found myself trying to work out which one of these toxically generous people was most likely to pay for me and how to thwart the devious philanthropists.

* 'Oh God, no, I couldn't possibly accept. Are you absolutely sure? I wouldn't want to …' yarda yarda.

Malaysia was like a sponge bath after the rugby match of Indonesia. English was spoken more widely, and the roads were smoother and less likely to snap towards the vertical. There had been no heckling, few horns, and nobody was insisting that I buy their wind chime. Was that relief I felt, or disappointment, or both? It was a more gentle and reflective ride, but it was a little lonelier, a little less exciting too.

Before Claire joined me, I'd grown accomplished at spending time alone, but now, with a conspicuous space beside me in a tent too big for one, I knew that I'd have to readjust. Solitude was not so difficult here, even comfortable at times, but it was much tougher in places where other travellers were mingling and having fun, though perhaps less fun than they appeared to be having, for loneliness warps your sense of the world.

Aloneness is a consistent trope in the chronicles of the early travelling cyclists. From the very first tours it was common practice to name your bicycle, a way of devising camaraderie where there was none. Some riders even talked to their bikes. Fred Birchmore was an American who cycled around the world in the 1930s. He became so numbed by months of solitude while in India that he stole a baby macaque from its mother, fed it buffalo and goat milk, named it 'Vociferous' and smuggled it past borders in his bread satchel. Of visiting the Taj Mahal he writes: 'As I looked at that beautiful gem in stone, the first tinge of homesickness touched me because there was no one save little Vociferous with whom to share the joy and beauty of these moments.'

Sadly, little Vociferous became markedly less so when she froze to death inside his satchel in the Burmese mountains.

*

In fading daylight, a trail tunnelled into the palms. I followed it until the rustle of fronds replaced the drone of traffic and the light turned wine-bottle green. Each footfall was like a landmine: one sent a monitor lizard bounding, the next unleashed a bat from a withered frond, and set the undergrowth rustling with fretful rodents. I was glad to see that an unnatural habitat could be so repossessed, but felt saddened too; this was just a temporary, salvaged home, and a reminder that the trilling, screeching strata of life in the forest was lost to the human appetite for more, more, more.

Camping among the palms, evenings began with 'sweat time'. Sweat time was non-negotiable and took around twenty minutes. In such supreme humidity, any activity, even cooking, meant a flash flood of sweat, so I'd lie down on my Therm-a-Rest mattress – as if on a sunbed, back first, then belly-down – before attempting anything else. I'd then have to mop myself with a truly upsetting thing I'd come to think of as 'the sweat-rag', and which might, if it got much sweatier, open up a porthole to hell.

Lying still, listening to the high song of mosquitoes, I began to cool down. I needed a pee less than I needed penis-malaria,* so I waited as long as I could. Even inside my tent, I was not alone. A cricket bounced about. A spider flickered in and out of polyester creases; beetles weaved past caterpillars on expeditions. Some of these insects had intruded through gaps in zips or perforations in my groundsheet, others might have been in my tent for months for all I knew, rolled up nightly. The caterpillar in particular looked familiar.

Malaysia, flat and palmy, was doing wonders for my mileage, but it was ruining my legs. Both of them were branded by a grisly rash that I was trying to get to the bottom of. I'd been stung by some nameless plant two days ago while traipsing

* Not a real thing.

through the palms; the sweat rash was now liberally spread and decorated by a spate of mosquito and horsefly bites. The chafing was outrageous, and behind my left knee there was a kind of pimpled, erythematous chaos that was looking more and more like the re-emergence of smallpox.

Even in the heat and closeness, sleep came effortlessly. Cycle dusk until dawn and you can snooze more or less anywhere, in any position, which was helpful because I had to. The seams in my air mattress had burst again, ballooning parts of it, so that I slept contorted, looking like the victim of a road accident. I drifted off to the sinister susurrations of palms, running scenarios: thief, police, leopard, cyclone … zzzzzzzz …

The next day I began to notice Indian-looking men with red bindis sitting on the edge of plantations outside plywood huts. Seven per cent of Malaysia's population are of Indian descent, reaching the peninsula in waves, the largest breaker landing with the expansion of empire when they were shipped in to work on plantations, roads, railway lines and ports. A few of the men grew curious, flagged me down and asked where I was from. My answer was met with 'cricket!' but we didn't have enough shared language for anything more, and grinning inanely at one another became a shallow, though pleasant, substitute.

The road winding up from Sungai Koyan was a new one and it would lift me through jungle to the Cameron Highlands before dropping me at the historical town of Ipoh, whose appeal for me lay mainly in the likelihood of it stocking palliative skin creams. But this topical paradise was still some way off, so I proceeded for three more days past more palm- and rubber-tree plantations. Then, finally, a snatch of something not yet jungle, but wilder – macaques scampering across a railway line and twirling one-handed from overhead cables. A sign warned of tapir, and more monitor lizards blustered across the road. A few young boys zipped past me on motorbikes, apparently on a break from breastfeeding, but generally I was alone, and soon I was smitten with the view over the forest canopy, a bounteous blend of palms, creepers, dead trunks, and lawn-green leaves.

Greenhouses appeared, like blisters on the hills, the land then folding into bright-green tea plantations before Tanah Rata. It was a town where travellers and tourists thrived, most of them with priorities at odds with my own. They were exploding into activity, their journeys just begun, shopping, hiking, sightseeing, renting bikes. Meanwhile, I began to triage my tasks: wash clothes, eat, wash myself, un-mat my hair, eat, fix my punctures, photograph my rashes, eat, wash food from my beard, hit the Gaviscon, mooch about (quietly groaning). Then off again, for more of the same in Penang.

I span off the highlands and by evening the last of the daylight was playing over the limestone hills near Ipoh. The following day I was looking out over a loading dock full of ship containers, the 13.5-km bridge that transects the Penang strait, and the white, tacky-looking 65-storey-high Komtar Tower, once the second-tallest building in Asia.

After four years on the road I was thoroughly broke, but scraping by thanks to the odd talk at an international school and a gentle stream of freelance journalism. I didn't have the luxury of pre-emptively changing bicycle parts so I rode until components disintegrated, leaving local mechanics astounded at the wear and tear, like doctors inspecting advanced pathology. Packet noodles, salty sacks of monosodium glutamate with the nutritional worth of talcum powder, had become something of a staple now. To save even more money, I'd begun to avoid cities and their attendant overheads entirely, or I'd sneak out of them come dusk and secrete myself in the edgelands, under bridges, beside railways, in the dank trimmings of car parks. If I was filthy or laggard or lonely, or otherwise in need of a hostel, then it had to be the cheapest around. In Penang, a guidebook assured me, this would be the Love Lane Inn. Incredibly, the Love Lane Inn was even seedier than it sounds, though it was not actually a brothel. The brothel was across the street, where prostitutes, of a surprising age range, some fluidly gendered, scattered every now and then on the alerting shriek from their pimp, which meant that the police were swinging by and with such enthusiasm that you began to wonder if it was for the good of the public or themselves.

The manager of the Love Lane Inn was Tan. Tan looked like Ozzy Osbourne if Black Sabbath had never split up and Ozzy had strengthened his commitment to stiff drinks and hard drugs. He had matchstick arms, an insalubrious pallor and, when he moved, he seemed to slink. Tan will stalk my nightmares for years to come.

The morning after checking into the Love Lane Inn I woke in my bunk bed to discover that the rash on my legs was now the least of my problems. Overnight, I had developed a biblical plague. Gross, scarlet, Old Testament weals had erupted on every inch of flesh, my fingers were bloated frankfurters, and the skin on my arms and shoulders looked like plucked chicken smashed repeatedly into a nest of fire ants and dunked in lava. I wanted to scratch more than I wanted to live.

Shuffling downstairs, I found Tan hunched by the front desk.

'Tan. I'm covered!'

He looked up, but without much interest.

'Fucking mosquitoes! Look, I'm covered in bites.'

'Nah. Not mosquitoes.'

'What is it then?'

He took a long pull on his fag.

'It's our bedbugs,' and he puffed out a grey cloud.

I had to remind myself that Tan was the manager of the Love Lane Inn, a fact that left me momentarily short for words.

'Our bedbugs?'

'Oh yeah. Man, we have a lot! Look. Everyone scratching.'

It was true. A whole posse of backpackers were sullenly grating themselves raw. One girl was scratching her friend, like chimpanzees grooming.

'Tan ...'

He was making an 'oh-what-fun' grin at the scrabbling travellers now.

'Can I change beds?'

'Well, you can, but ...' He made a deflating sound.

'You can't hide from our bedbugs! Man, they're everywhere! It's amazing.'

'It ... it *is* amazing,' I said, referring to Tan's apathy to deal with it as much as the scale of the infestation. When hostel managers are challenged by a guest raving about bedbugs, they tend to blame something else, anything else, because bedbugs are tenacious little fuckers, and dealing with them often means closing the hostel, an arduous decontamination and a hunk of lost income. Tan seemed unlikely to walk this road. Almost every cheap hostel in Penang was crawling, and if one or two managed the infestation, the well-nibbled travellers from the others would check into them thus reinfecting the beds, which meant that the brothel over the road was potentially the least disease-ridden place in town.

I checked out of the Love Lane Inn – Tan managed to look puzzled by this – and struck out for George Town's night market, which was arranged on each side of a busy road so that customers queued for food amid a ferment of motorcycles, rickshaws and cars. Women cawed instructions to table runners, who shimmied and bobbed about the traffic as pedestrians huddled in conferences about what to eat. Tentacles of people reached into the street behind tented carts where woks spat and cracked. The vendors were one-man-bands of the culinary craft – tossing, throwing, frying and chopping, each action nimbly blending with the next, puffs of steam greying the night air, the cooks emerging now and then, like magicians.

I bought dumplings, stabbed at one with a chopstick and brought it to my mouth. It tumbled and plopped heavily into chilli sauce, which splashed into *both* of my eyes. The pain was instant. Tears poured down my face. Hungry and half-blind, ravaged by parasites and unable to reliably seek out a clean hostel and inspect the beds for lice, I made my way back to the Love Lane Inn where Tan was still gently dying.

'You want your bed back?'

'Yes please, Tan.'

I set off the next day towards the Thai border, and Thailand, I'd assumed, would be a cinch. Flat, at least to begin with, lush forest, cheap hostels and heaps of *pad thai* served by a folksy band of smiling, bowing, nice people. The resort town of Ao Nang, however, killed the cliché. I checked into a hostel dorm and woke in the small hours to a pulsating pain in my head, a kind that made me want to trepan my own skull with a bicycle chain tool. I had a fever now too, an unrelenting, 39-degree sweat-a-thon, broken by little patches of disorientation. The next day was a write-off. I suffered in bed until it blended inappreciably into night and then got up, unsure why, unsettled by the logs of bodies around me and a new agony behind my eyes.

By morning, an army of papules had spread out over my abdomen. With the arrival of pain in my molars and elbows, everything now hurt, but not quite enough to distract me from insatiable thirst, some diarrhoea (memorably profuse, profusely memorable) and the disturbing fact that my gums had started to bleed. Mentally, I sorted through the list of suspects, and with some dismay settled, correctly it would turn out, on dengue fever.

Dengue is a virus, transmitted by mosquitoes, and a pretty new kid on the block, probably moving from monkeys to humans less than 800 years ago. It has an apt nickname, 'break-bone fever', and it makes you feel all-over-shit. Most people recover without complications, though it was my uncharacteristic inability to eat that seemed to declare my likelihood to suffer all of them. In a small medical clinic, a Chinese doctor took blood tests and delivered the news. 'Dengue!' he said, using the tone of a darts commentator. 'Oh yeah. Is bad luck for you.'

Luck, huh. Luck was having less and less to do with it. Dengue is ever more generously bestowed by mosquitoes and a chart of its incidence over time looks something like a skateboarding ramp. Still, this was incongruous. Thailand was supposed to be the land of smiles, not the land of devastating tropical disease. Pallid, sweaty people with bleeding gums and handlebar ribs don't stare dead-eyed from the brochures. There are no slogans: 'Break-bone Fever! Everything it's cracked up to be!'

By the fourth day my fever had broken, but, impossibly, I felt worse than before. I plodded down to the lab, where the doctor related my blood test results and we glumly appraised the latest trench in my white blood cell and platelet count. The plunge continued into the fifth day of my illness, the least fun period of the dengue infection, when you're most likely to haemorrhage your socks off. My mind filled with the photos of cases in medical textbooks, of puffy men bleeding around tubes stuffed into every opening, pasty and ventilated and looking as though all that treatment wasn't going to be worth the effort.*

I emailed my test results to friends from medical school.

* The Rule of Five – if more than five of the patient's orifices are obscured by tubing, he/she has no chance.

Ian was now one of Australia's highest-flying doctors in Queensland and Tom was an infectious disease specialist in Manchester. Initially, they both responded with 'shit!', which I didn't think was a particularly professional response for so-called experts in their fields.

In the medical tomes, dengue has a long list of symptoms of which I was approaching a full house: I was a clammy, tachycardic, anorexic textbook case. I had another symptom that had been curiously omitted from the texts – an irrepressible yearning to inform everyone that you have it.

'Try and eat something,' encouraged Martin, a German cycle tourer staying at the same hostel, who'd been informed of my infection, by me, immediately after I'd introduced myself.

'I don't feel like eating, Martin, on account of my dengue fever.'

'Okay, well do you need anything? Something from the pharmacy?'

'I'll try and go myself. Though I have dengue fever, Martin. It's tough, you know?'

Soon, everyone knew. I was careful to groan when alternating pillows. I wished for some more physically obvious manifestations of my pestilence, but the rash would have to do and I made sure to go about bare-chested.

Nobody knows for sure how dengue got its name, but I'm aware of three theories. One is that it came from a Swahili phrase 'ka-dinga pepo': an attack by an evil spirit. Another goes that West Indian slaves with dengue were crippled with pain and developed a 'dandified' way of walking, so dengue comes from dandy. The third is that it comes from a Spanish word that translates as *fastidious*, referring to the fact that patients with dengue were not inclined to walk anywhere.

After ten days I was back on my bike, and the first miles were hard won, but I woke the following day feeling a world better and – ta da! – my appetite had returned with a vengeance. I pedalled up the western side of Thailand, past wooden shacks stacked with pineapple, mango and papaya that were run by thin kids who stewed in hammocks nearby until roused by the whistle of a customer. Here and there I rolled past the hum and jostle of markets where the air was perfumed with barbecued meat, and where I tried to eat back the weight I'd lost to my life-threatening bout of devastating dengue-plague.

The heat took a toll and afternoons were a tipsy fight against the temptation to stop and nap for an hour or two. Between villages women wandered by in pools of pink and blue light, the sun burning through the shelter of their umbrellas. In the late afternoon a small group closed in around me, a ring of sudden, brown-toothed grins. One looked ancient, her left eye fogged by a cataract. She began bawling at me, benevolently and in Thai, which tripped a wave of laughter from the others. A younger woman with some grasp of English chipped in, gabbing pell-mell through different topics: 'Are people friendly in England? No. They work too hard! Thai people like blue eyes, light skin. But white people like dark skin! Have a pineapple!'

The others whipped out guavas and custard apples in a quick contest of giving before wandering off chattily down the road.

In Ranong, I sat in a restaurant, puzzling over a poster in which Coca-Cola purported to be official sponsors of Ramadan. I'd always assumed that Coke had a monopoly on Christmas, and as I mused on this act of corporate apostasy, some murmurs arose from near the TV. A banner in English began to scroll across the lower portion of the screen: 'Following the implementation of martial law the following are appointed …' It continued with a list of military personnel, and then the words 'National Peace and Order Maintaining Council'.

Take that, Orwell.

I played out a brief phone conversation in my mind: 'Yes, Mum, martial law ... yes, a violent revolution ... Yes, of course I'll be fine. I hear there's a helipad on the roof of the British embassy ...'

Though martial law and a nationwide curfew had been declared, General Prayuth Chan-ocha, Commander of the Royal Thai Army and a rather diffident figure, I thought, for a coup leader, was keen to reassure everyone that this was not a coup, only to change his mind two days later, ordering the immediate detention of political players and journalists. For the BBC, whether or not there was bloodletting was neither here nor there; what mattered was how this was going to affect the travel plans of all those poor, virtuous British tourists, so very far from home. Sex tourism never had it so tough.

Dissent, generally, was tepid. Students gave out 'sandwiches for democracy' on the streets of Bangkok or made three-finger salutes in a gesture borrowed from *The Hunger Games*. Soldiers held bouquets of flowers and posed for photos with passers-by. 'Martial-law selfies', as one newspaper described them, were a thing. But then this was run-of-the-mill in Thailand where there have been nineteen coup attempts since 1932. They roll around only slightly less often than an election.

I picked up a newspaper to read that the Thai army had begun a surreal campaign of 'bringing happiness' to Thailand, which involved a variety of festivals and shindigs, free food, health checks and haircuts. Ironically, it was populist manoeuvres of the former government that were cited as a defence of the coup in the first place.

With the Thai constitution suspended and the curfew still in effect, I took the highway north to Bangkok, hoping society didn't collapse along the way. At one point, Thailand thins out to only 450 metres wide, a slip of a country between Myanmar and the sea, and as I followed the canals north, a saw of Burmese peaks lifted slowly in the distance. I stopped just once that afternoon for a soft drink. In the shop's fridge, I noticed two dogs, nestled between cans of Pepsi. I stalled. Must be kids' toys, I decided, but the realism was jarring, the wet noses, the veiny ears ... I eased open the door. The dogs stirred and peered coolly up at me. I reached behind them for a Pepsi, pulled the door to and handed over money to a woman at the counter, who'd sensed my alarm.

'Thailand hot hot hot! Better my dogs in fridge!' she said.

I later confided this story to a British friend living in Thailand who exploded into laughter and said simply 'Well, Thai people can be very practical'.

*

Bangkok is hot, really hot, the hottest major city in the world in fact, if you look at annual averages. It's hot on traffic too, after a long policy of converting footpaths into roads to ease congestion and a long history of politicians selling off public policy to the highest bidder. It's not a city intended for pedestrians but a place that makes you yearn for a future in which cars are exiled from city centres and bike lanes and pavements are spacious, a future from which we'll look back on this grim, noisy era with disdain, as we do the stench and grime of the Industrial Revolution today.

The pavements in Bangkok are scattered with manholes into which people regularly disappear screaming – an entire taskforce was once set up to counter the issue. Streets are commandeered by mobile dishwashers, customers spilling out of eating houses, and by motorcycle dealerships, loading docks, election advertising, arc welding. If you hit a clear patch, the motorcyclists will have noticed it too and you'll have to share. Street signs are held up by steel hawser wires, attached to the pavement. Guttering empties onto all of it. Electric cables dowsed in this run-off gently fizz. It's actually a rather thrilling experience.

Not so much the Khao San Road. I knew I was getting close because the pantaloons on travellers were becoming increasingly dramatic to the point that MC Hammer himself would have looked only aspirationally baggy-legged. The famed 'backpacker zoo' was upon me, heralded by an explosion of signs from the side of every building, like the outstretched arms of beggars, or prisoners behind bars. A girl was getting hair braids with a look of uncertainty. Bookstores were loaded with copies of *Shantaram*, with whole sections devoted to 'mind drugs'. Tuk-tuk drivers heckled for custom and vendors touted the same vests, this chunk of Bangkok neatly summed up by their slogans: 'same shit anywhere'.

Being footloose was not a quality often encouraged on the Khao San Road and every two minutes I was forced to defend my aimlessness. 'What you want? Where you wanna go?' I decided that I wanted a museum, which I assumed would be less stressful. Unfortunately, I chose Bangkok's museum of forensic pathology.

In the first display I came to, there was a line of photos revealing some unfortunate, very dead people. A man decapitated in a train wreck, his severed head left inches above his torso on bloodied sheets. I admired murder victims, finished off by knife wounds and bullets. I was not grateful to learn what a hammer attack might look like, or the bloody consequences of a sword fight. One sign read 'throat cut by beer bottle'; my eyes inched up to the image – yep, I never want my throat cut by a beer bottle.

I supposed I was lured here by my general fascination with life, death, biology and bodies, but I wondered now whether the curator's intention was to educate or to shock visitors into a state of permanent reclusion. This wasn't the mind-expanding family day out I'd assumed most museums to be, especially as in another room there were stillborn infants, deformed beyond the fancies of science fiction, and in another the mummified remains of the city's most enthusiastic rapists and murderers who'd been sentenced to death. There was a parasite room too: centre stage, the prize exhibit was the scrotum of a man with elephantiasis, and it was so enlarged that you could easily squat inside, should the desire overtake you.

The Gordon Museum of Pathology at King's College had long been one of my favourite ways to pass the time in London. It's only open to medics, because of regulations within the Human Tissue Act, but if you want to see what a necrotic brain tumour looks like, it won't let you down. There are jars of unspeakable things. There are foetuses with cyclopia (no nose and a single central eye, like the cyclops. Such malformations are the result of a defect in the Sonic Hedgehog Gene Regulator – so-named because the defective version of the gene in fruit flies causes the embryos to be covered in small denticles, like hedgehog spines. Never let it be said that molecular geneticists have no sense of humour). There are foetuses with diprosopus too: literally 'two faces' (and, literally, two faces!). There is even a section dedicated to infanticide.

There will always be some blurring of the line between distasteful voyeurism and genuine scientific interest, but for the curators of the museum in Bangkok, only the photo of a woman who had been beaten to death by a dildo had been deemed overkill, if you'll pardon the expression, and had therefore been removed (though whether for its gruesomeness or its literal and metaphorical misogyny I wasn't sure).

In Bangkok I was staying with Elena, a friend who lived behind a string of bars where bands of rough-looking men jeered at kick-boxing on TV. To get to Elena's house I was reliant on the city's motorbike taxis and my driver was a hothead who apparently believed that the hour of his death had been established before birth and that his driving could not influence it. I wanted to quibble, to stammer something about his speed, but I worried that this would be construed as impolite in Thailand so I hugged him silently, disgusted both by his recklessness and my own readiness to face death for the sake of an awkward social situation. Meanwhile, photos of road-accident victims flashed back to me in gory Technicolor, complete with glassy eyes and captions: 'Lacerated liver'; 'Tire tread marks on abdomen'.

We arrived, and I embraced Elena with all my (lovely! intact!) nervous system, then packed a bag and pondered a detour.

8
Pulse

I spent a few days in Bangkok poring over maps of South East Asia, fantasising about all the places that nobody had told me to go. I *had* planned to move west, but maps are dangerous to plans and a siren call to adventure. My eyes get drawn to the small, shy roads, the unmarked spaces, the towns with strange or playful names. Through maps, I stray, and my life is thrown off course.

Today, I was lost in the Tonle Sap, a blue splodge in Cambodia, the largest freshwater lake in South East Asia. Although not indicated on my map, I'd heard that the lake was scattered with villages made up of floating houseboats. Photos online suggested a pastoral Venice: men, standing up, shifting sampans with wooden poles, their arms muscled by years at the task. At least a million *neak tonle*, or river people, were said to live on the floodplain. I had a friend, Ian, a British family doctor living in Cambodia, who'd mentioned a floating medical clinic on the lake. I decided to leave my bike in Bangkok and pay a visit.

This was more than a detour in the physical sense; larger ambitions were changing course too. By now, the slow task of biking was wearing thin. I felt myself growing self-absorbed, and wondered if I was heading towards something like burnout. Travelling still helped me pose questions about the world, but it didn't often answer them because I was only ever a fleeting visitor, without the language or immersion to get to the heart of things. The mobile clinic in Turkana and the dump in Jakarta spoke of an underlying interconnectedness that fascinated me. I wanted to see more of hospitals, clinics and healthcare projects on my way home, to indulge my curiosity if nothing else. I would be nowhere long enough to work or to volunteer, but in sniffing around and writing – perhaps 'journalism' was too strong a word – I might discover a fresh sense of purpose. Two journeys then. The physical ride was the classic traveller's 'quest for anomaly', the sensory rush, the mud and wind, the adventure. But in parallel would be a journey to understand something of the forces shaping health, and, by proxy, places themselves. This, after all, was how London had been best revealed to me, not simply through residing there, but through the NHS, which led me to Londoners of every stripe and gave me a new angle on the city. And for this parallel journey, too, my bicycle would do nicely. A travelling cyclist is often invited into homes and communities, where the undercurrents of health and disease can be easier to see.

*

I took a bus across the border into Cambodia and changed into a rusty truck that dropped me at the village of Kampong Khleang, a muddy synapse of landlubbers and *neak tonle*, with homes stilted over greenish-brown water. The most extraordinary quirk of the lake resides in its rhythm. In June, the Tonle Sap drains into the Mekong, but the direction of flow in the river switches twice a year, and by October the water backs up, sinking forests and the lake swells to five times its size. They call this seasonal see-saw the flood-pulse.

With an invitation to join the medical team for a few days, I sat on a small motorboat alongside a couple of doctors, a dentist, some nurses and a cook. It didn't take long to see that the Tonle Sap was turning into a marsh. Blame a prolonged drought, dams on the Mekong (there were more than a hundred being knocked up on its tributaries) and water hyacinth, an ecological plague that can grow up to five metres a day, glutting the surface, stealing oxygen, blocking waterways and making whole villages unreachable. When French explorer Henri Mouhot crossed the Tonle Sap to 'discover' Angkor Wat (I'm guessing that, like most western adventurers in the mid-nineteenth century, he left off the inverted commas) there was no hyacinth, and Mouhot wrote of the sheer numbers of fish striking oars and the hull of his boat. Overfishing too has helped forge an ecology that the ancient peoples of Angkor Wat would not recognise. We cut through the water, our hull untroubled by fish.

After three hours the waterway tapered through mangroves; a few motorboats, laden with fish, left chevrons of ripples. As the channel broadened, the floating homes appeared, roofed in crinkled zinc or thatch. Some had submerged metal tanks for floatation, others, boat hulls and bamboo. Picharkrei: home to some forty or fifty families and a hut flying the white flag of the Lake Clinic.

Many of the villages on the Tonle Sap didn't simply move up and down with the annual flood, they shifted position too, the houseboats flocking to a new spot. How do you define a village? Can you bestow a name if it has no steady location? Perhaps a name relates to the community, or the buildings, not the place. But on the Tonle Sap these were in flux too. People were drawn to seasonal work in cities or to marry, or to the paddies outlining the lake to pick rice. The houseboats were rebuilt and added to. Naming these villages seemed to me like naming a wind or a sand dune or, I supposed, a person: after all, humans are in a constant state of transition, replacing cells, shifting character, perspective and even sense of self. We are not who we were a day, a month, a year ago – though as

Alexei Sayle points out, this is not something Thames Water will accept as a reason not to pay your water bill. I thought about how philosophers have long speculated on the transitional state of human being. In *A Treatise of Human Nature*, Hume wrote that we are 'nothing but a bundle or collection of different perceptions, which succeed each other with an inconceivable rapidity, and are in a perpetual flux and movement'. I agreed with Alan Watts's assertion too, that the conception of ourselves as 'an ego in a bag of skin' is not supported by science or intuition. More than simply being false, it struck me as dangerous to perceive ourselves as immutable beings. Perhaps mutability helps us thrive together in a riotous world. Without it, where's the hope that we can?

*

Inside the airy, metal clinic, we readied for the coming patients, drawn from seven floating villages nearby. Bodies slowly filled up the benches until the clinic was part waiting room, part consultation room, as well as kids' playroom and village hall, and cacophonous with what I fancied was juicy village gossip. The villagers were a mix of Khmer and Vietnamese, and a small number of Cham, a Muslim minority. A few had never been to the mainland, though not many of the adults were born on the lake – some had been pushed here through debt and poverty, and many had lost their homes and relocated to the lake following the civil war.

There were no heroics here, just simple healthcare: education, prevention, screening, dentistry. I began the day with Savaan, an avuncular Cambodian nurse, flitting between the first to arrive, taking vitals on the clinic's porch and triaging the sickest. Savaan had an easy way with his patients and kids rarely escaped without a ruffle of their hair.

I hadn't imagined treating patients myself, helping thin, doe-eyed Cambodian kids for a week before cruising away for some sightseeing. I couldn't dunk myself into this world, take time to get to know its bones and its spirit. I'd planned simply to observe, but I soon realised I wasn't going to get away with that, and I was pulled into consultations by the nurses. So, cautiously, and with plenty of help, I began seeing patients myself, or advising a little where I could. Occasionally other foreign doctors came here to volunteer, and often they were well-trained and newly garnished with diplomas in tropical medicine. But, like me, they were grossly disadvantaged in not knowing the *neak tonle*, the veiled pressures on their health, and, vitally, the language, even if the nurses acted as translators. What, I wondered, were these small marks, evenly spaced and circular, like cigarette burns, on one woman's arm. I asked Savaan.

'Traditional healers! You see, they make little marks like this. Sometimes they get infected. I try to tell people not to go.' He waggled a finger in the air, like an incensed professor.

With one doctor for every 10,000 people, Cambodia had one of the lowest ratios in the world. Almost every country below it was, predictably, African. But nature abhors a vacuum, so alongside healers, people went to unlicensed doctors. Many were former military field medics with basic knowledge and training, and they often usefully filled the gap, but they'd been known to cause problems too. Intravenous fluid was a popular therapy, considered a potent cure for all sorts of things. In this way, babies in particular could get overloaded, their lungs swamped.

Everyone knew the story of Yem Chrin, a war veteran and unlicensed doctor, twenty years in the game. He'd volunteered in the refugee camps along the Thai border during the days of the conflict before settling in a village called Roka in

Cambodia's north-western Battambang province. For decades, he reused syringes on his village rounds, inadvertently infecting his patients with HIV. The number varies, but most estimates put it at two hundred, from six-year-olds to seventy-year-olds, including celibate Buddhist monks and sixteen members of one extended family. Twelve per cent of those eventually tested in Roka turned out to be HIV positive, thirty times the Cambodian average.

Yem Chrin was revered. His treatments were cheap and popular, and sometimes given on tick or in exchange for household goods. As the disaster surfaced, the Cambodian prime minister rejected the theory out of hand, blaming the tests, but as more international researchers got involved, there could be no refuting the results. Yem Chrin was jailed for 25 years and the name Roka grew ugly on the lips.

Two teenage girls sat down in front of me. Savaan asked them what was wrong. They looked a little anxious at the question before a brief tête-à-tête in Khmer. Headache, they decided. Both of you? Yes. A friend popped up behind them. And you? Headache too. Right. This wasn't the first time I'd wondered if villagers had come to the clinic simply because we were here, giving stuff out.

The clinic was now full of women whose loose, patterned clothes carried in the scent of woodsmoke. There was the usual gamut of family-practice patients: teenagers gloomy about their acne, older ladies shuffling in with painful arthritis. Others came with ailments bound up with their lifestyle on the lake – a fish smoker with a cough, babies with diarrhoea and eye infections. Intestinal parasites were rife.

There were others, too. A grubby boy clambered onto the chair, lost inside a Manchester United top. His mum, her face fretted with wrinkles, sat down beside him. The pair looked forlorn, uneasy and very poor. 'My son sounds like a cat,' translated the nurse, which I retranslated, without my stethoscope, to mean wheezy. I'd heard him from the bench. The boy stepped onto the scales – the needle swung to fifteen kilograms, less than I'd hoped for a nine-year-old. 'Skinny, dirty …' said the Khmer doctor, as if announcing the futility of a few vitamin pills when there were forces at work well beyond our ability to set right. I'd felt a similar impotence before: you come into medicine with an understanding that disease and pathology are the threats to health, but the more patients you treat, the more you see that it's the conditions for disease that are the real enemy. Saving lives might be the romantic notion would-be doctors fall for, but the reality can feel more like firefighting.

For the men, a visit to the Lake Clinic meant time off fishing, and I began to steel myself as we examined the ones who did show up; their ailments tended to be long-standing and more severe than the rest. I noticed one man in particular, who sat alone in the waiting room, which was unusual in itself. The lump in his neck made dispensable the usual question of how we could help. I palpated it gently: it was fixed and firm, there was no pulse, no pain to my touch. His raspy voice spoke for its effect on the vocal cords. There were no surgeons, scans or biopsies out here, and if it was cancerous, he'd need more than fish to pay for treatment. The Khmer doctor explained the problem and the man didn't flinch. Perhaps, I thought, the misfortune of poverty toughens you against the capacity to feel more, but as I watched him pause on the threshold of the clinic and stare dismally into space, I saw that this was my mistake, my bias. Death, they say, is democratic – but not a good death. Not everyone has that opportunity.

*

I returned to Bangkok and by the following afternoon I was

riding north-west on a lacework of small roads beside rice paddies. Trucks passed, loaded with freshly felled sugar cane, tailing a sweet-scented wind. After a few days, the horizon lifted into smoky mountains and that afternoon a young Chinese man pulled alongside me on a red mountain bike. Chieng had none of the reserved manner I associate with the Chinese. Perhaps he was naturally exuberant, or perhaps he was allowing himself to blow off more steam than usual, for this was his first journey outside China. Excitedly, Chieng told me of a free coffee he was given at a police station two nights ago and I smiled at the simple things that mean so much when you've pedalled 150 kilometres.

As we sat spooning up *pad thai*, I asked Chieng about his life in China. He held up a defiant hand.

'I love my country but I hate the party. We have no freedom! We ...'

A chorus of beeps stopped him dead. Chieng scowled at his phone.

'My mother. She always ring me. Every day.'

He picked up the phone but held it at arm's length, frowning, like it was a bad exam result.

'I want to cycle around the world too,' he said, dreamily. 'Chinese parents ... they don't understand.'

Back on the highway, I could see that Chieng had a more pressing issue than his cossetting mum. He was spooked by dogs and the loops he pulled around even the meekest of creatures were absurd. I showed him how to go on the offensive by lobbing stones and I was immensely glad to see him launch his own attacks. I'd recruited, indoctrinated and trained another combatant in my inter-species jihad.

We parted ways soon after that, a sign steering me away from another fleeting friendship and towards the Myanmar Friendship Bridge. Chieng swung right with a long-chinned grin and a wave, his heart set on Chiang Mai, and beyond, if his mum would let him.

Chieng was one of hundreds of cycle tourers that I would meet on my journey, a great sprawling family of teenagers and septuagenarians, lightweight 'credit-card tourers', men in spandex body suits and raggedy bike-packers on fat bikes. Some were on recumbents or tandems, a few tugging trailers with a toddler or even a dog inside. They'd been away from home for three weeks or five years. We wished each other tail-winds. I felt a tribal loyalty, an *esprit de corps* – we shared the same rituals in camping wild, shitting in the woods, endless eating. We had the same enemies in officialdom, weather, ruth-less dogs and lunatic traffic.

People have been 'cycle touring', which is to say travelling for adventure, pleasure and autonomy, since the tail end of the nineteenth century, not long after the bicycle craze began. Cycling was advocated as a remedy for a variety of ancient-sounding ailments: torpid liver, incipient consumption, dyspepsia, nervous exhaustion, rheumatism and melancholia. It was alleged to make teenagers grow faster. Later, cycling was proposed not just as a tonic but as a way to explore places few others did. There was a genuine sense of ecstasy in the writings of the first cycle tourists. Is it just me, or are things getting a bit racy?

> Touring is the backbone of cycling and nothing can surpass the pleasure and exhilaration of wandering free as the wind o'er hills and dale, with no care and anxiety but to drink in the fresh air and sunshine, to feast the eyes on natures beauties and feel in every vein and sinew the throbbing and power of vigorous manhood.*

* R. J. Mecredy and G. Stoney, *The Art and Pastime of Cycling*, 1895

Actually, back then cycle touring was more like bike-packing, its venturesome cousin, which more often involves dirt roads, minimal kit and frame bags. The bicycle drew special attention to some Victorian adventurers and had all the gimmickry of a stand-up paddleboard today. There was a whole new set of distance and speed records to be set by bicycle. The earliest cycle tourers had various incentives for their journeys, fame and fortune included. One of the first was John Foster Fraser, a staunch imperialist who undertook the longest ever bike ride at the time by riding 19,237 miles around the planet from 1896 to 1898, alongside companions Samuel Lunn and Frances Lowe. This was at the peak of empire, when a quarter of all the land on earth and a quarter of all its peoples fell under British rule.

Fraser is quite candid about his ambitions in the opening of his book, *Round the World on a Wheel*, published in 1899: 'We took this trip round the world on bicycles because we are more or less conceited, like to be talked about, and see our names in the newspapers.'

Wearing 'brown woollen garb, guaranteed by the tailor to wear for ever and a fortnight' and big bell-shaped helmets, they departed Britain amid much fanfare. Rampant self-publicists, they were applauded in the papers as they crossed Europe. In Belgium they were granted an audience with the king and in Vienna and Hungary they were courted by the press. Union Jacks were frequently hoisted in their honour, journalists were immediately summoned and bands conscripted to entertain them with British music-hall tunes. They dined and danced with Transylvanian nobility.

There is, of course, something of the Victorian age in their sense of superiority, and some of their descriptions make your eyes water. In a Turkish village: 'The whole populace surrounded us: sore-eyed old reprobates, unshaven ruffians, meddlesome young rascals, inquisitive women, about two hundred in all, gaping wonderstruck at our worthy persons and machines.'

To get some sense of how wealthy Europeans travelled at the time, or for giggles, it's worth consulting Baedeker's *Traveller's Manual of Conversation in Four Languages*. Here, you will find translations for 'I want my bed warmed directly', 'I am suffocated with rage', 'If you do everything to my satisfaction, I will reward you liberally', and 'Have you fresh leeches? These do not bite.'

Fraser and friends were so arrogantly pumped by the simple fact of their nationality that when they got into trouble they seized their British passports and waved them in people's faces, and yet they were a lot more reliant on local hospitality than I had been: they carried no tents and no stoves, and demanded that local people host them, feed them, congratulate them and do their bidding. 'The only way to get a Persian to do anything, and to do it quickly, is to ride an exceedingly high horse', wrote Fraser.

In Persia, an old man told Fraser: 'You have the love of all our people, and tomorrow morning when you leave us you will take with you a hundred muleloads of hearts!'

Fraser declared Persian villagers 'pucker-faced and scraggy'.

*

I was pucker-faced now as I rated a haze on the horizon. Weather in Thailand often started like this: fast and from the side. In minutes, the road turned to fizzing puddles, and I was soaked through as I climbed into the Tanonthongchai mountains towards the border town of Mae Sot.

I had a contact at the Shoklo Malaria Research Unit (SMRU) here, a gang of medics and health researchers who'd been treating migrants and refugees living on both sides of the border for almost thirty years. Since the 1970s the border between Thailand and Myanmar had seen waves of refugees who were now settled in camps on the Thai side. More Burmese had chanced India, Bangladesh and Malaysia, but 160,000 refugees were officially recognised here. The Thai army had taken over command of the camps near Mae Sot a few days ago, evicting staff from NGOs who were tasked with helping the refugees, and ordering searches on the pretext of 'drug enforcement' (read: uncertified people). It had been swift and easy to do so as there was still no government in Thailand to answer to.

It was a blue and sedate day when I met the Burmese doctor, Dr Myint, outside a TB field hospital set on a broad hill not far from Mae Sot. She had a cool seen-it-all-before manner, earned in less peaceful circumstances, having worked with Médecins Sans Frontières in Somalia and other battlegrounds. The view from the clinic was lovely and green, tapioca crops tumbling away on every side. Near an administrative building were three long wards, rattan thatch dividing the small rooms, one for each of the 76 patients. Dr Myint rounded up a couple of nurses and we trotted off on a ward round.

She eased open the door and there was our first patient: a figure lying on the ground. A woman, but not obviously. She was emaciated. Her head was a dark skull, the skin stretched taut over cavernous temples, and her hair had fallen out. A bloody crust encircled her mouth while a tube disappeared into her nose, attached to a bag of saline. Her bony chest shivered with each fast breath, and her eyes roamed, large and nervous; otherwise, she was still. Her name was Paw Htoo, which was about the only personal detail anybody knew. My view pulled back now. There was a man too, squatted beside her, stubbly-headed and well built. In his two large hands, he held one of hers.

Dr Myint knelt on the floor beside her and spoke softly in Burmese. Paw Htoo's mouth twitched but it was a fruitless attempt to form words. This was palliation – and if that wasn't clear from Paw Htoo's wasted body, it was clear from Dr Myint's mien, and from the very air which smelt, already, of death. I knew that smell from intensive care, from leaning over patients in their final days with my stethoscope, listening to racing hearts and crackling lungs. The salty, sugary smell of time running out. The final seconds in that room were saved for silence. Dr Myint bowed and we left.

I'd been on ward rounds before when the team gets sucker-punched by a sadness. The ward round continues as it must, but mechanically. Outside, Dr Myint pursed her lips and said nothing. I noticed some plastic bags near the door.

'Yesterday, I found her with one of those over her head,' said Dr Myint. We were all looking towards the bags now, all of us made to see how futile the attempt at suffocation would have been. The bags were riddled with holes.

If Myanmar – international pariah, governed and sanctioned into a crumbling state – had suffered under the junta, then the ethnic minorities in the horseshoe of hilly borderlands had suffered even more. Paw Htoo was one of the Karen ethnic minority living in the hills of eastern Myanmar, where rebel groups had been in a long struggle with the Burmese army.

For Paw Htoo, HIV had precipitated tuberculosis and lymphoma, a form of cancer affecting white blood cells. Her CD4 T-cells, essential for immunity and the route via which HIV invades (and which in healthy people should amount to over one thousand per cubic millimetre of blood), had plummeted to three. Two weeks ago, she'd been left at the gates of a monastery inside Kayin State, just over the border, in Myanmar. Buddhist monks are forbidden physical contact with women so none of the men could carry her inside, swat away her flies, or feed her. One monk offered to remove his saffron cowl and leave the monkhood. He'd carried her here as she drifted in and out of consciousness and he remained by her side now.

The monk knew, as everyone did, that there was more hope on this side of the border. Hospitals inside Myanmar were often shabby, dirty places that offered little reprieve to those who needed drugs they couldn't afford and care the government wouldn't provide. In 2000, as Paw Htoo reached womanhood, Burma's healthcare system was ranked 190 out of 191 nations by the WHO. By most estimates, the junta (or 'The State Peace and Development Council' – their self-proclaimed and rather Orwellian moniker) were still not seriously investing in healthcare, which, at the time of writing, received less than 2 per cent of the national budget. At least 40 per cent went to the military. In a seeming corroboration of this fact, as I stepped outside the field hospital to take a breath without my mask, there was a rumble. I looked up. Above Myanmar, west of the mingled peace and pain and slow-death of our hill, two fighter jets split the sky.

The fly-by came early, but only just. Paw Htoo died an hour later. Nobody knew if she had family or friends somewhere who would need to be notified of her death. That afternoon, I saw the monk walking with his head down through the tapioca, making his slow way, alone, back to the river.

Officially, Paw Htoo's death would be one more for TB and HIV, diseases that, in the modern world, might not be so ravaging if your circumstances are the right ones. We've had effective combination therapy for TB for half a century now, and though a diagnosis of HIV in the UK today remains seismic for anyone, with contemporary anti-retroviral drugs your life expectancy will be close to normal. On death certificates, doctors are asked to state the 'underlying cause of death', and you might include the biological background to a person's demise; a heart condition or an infection, for example. You do not mention poverty, a rickety healthcare system or military rule, though it occurred to me that these were modes of death too.

*

The following day, I cycled beneath a white arch, its gilded trim twisting skyward, gold script declaring 'The Republic of the Union of Myanmar'. Beyond was a street lined with balconied buildings, where satellite dishes on walls all angled towards Thailand, like open ears.

Down the road was Kawkareik, a small town of dust and nervous dogs. Spidery men pedalled trishaws or sat in the shade of teak, leaf-roofed huts, bare-chested, dragon tattoos from shoulder blades to smalls of back. A policeman approached me, stinking of liquor and sending a red jet of betel-nut paste to the ground.

'I take you to guesthouse. Only place you stay,' he said, and escorted me to a large wooden house climbing with plants. The owner was smiling before I was in the door. I began filling in the guestbook.

'Ah you are September born!' he said, as I scribbled my date of birth. 'You are gentle, smooth and relaxed. You don't like to influence others.'

I gave the policeman a little 'have some of that' look and sauntered off.

The plywood-panelled room was just big enough for a bed, and the mosquito net had holes that could accommodate a pigeon. I tied up the holes as best I could and lay down, road-weak, letting my legs stretch long, and then drifted off to the sound of fruit sellers bawling for customers, the hum of tuk-tuks, dogs baying and scrapping. The first night in a new country is often the best.

The next day I continued west, and by nightfall, near Hpa-an, I was ducking behind some bushes to conceal my campsite. Inside my tent, I could hear men's voices. They came and went and popped up close by. Footfall too: slow, determined steps. Psychopathic steps. All went quiet again until a dog stalked over, a Doberman, probably, rabid and toothy, with yellow eyes. It had just the bark.

This primal sense of dread was one of many reasons I'd fallen in love with camping stealthily on the edge of towns. Sleeping in this liminal space can feel as heady as camping in the wild, especially when it's against the law, as it was in Myanmar. There was a feeling – during these nightly detours – of stalking society. It felt thrillingly outsiderish: I was the thief at the window. Childishly fun, like a game of hide and seek.

I was rough sleeping around the world as much as cycling around it: both defined my journey. Usually, it was easy enough to find a patch of earth to call my own, but I fancied myself an expert now. To dodge detection, I was strategic. Camping high beats camping low, for instance (people look down at things more often than up at them). Pitching your tent is best done on the cusp of darkness too, when you can evaluate your campsite before night falls and hides you away. Camp too early and you risk getting spotted. (This is worse than it sounds. In sleepy villages all over the world, you're news. One curious soul will multiply to twenty. Between eight and ten of them will be content to watch you sleep.) Camp too late, when it's dark, and you'll inevitably wake up the next morning beside a sign saying something like 'Warning. Mine clearance in progress'.

Of all the nights I'd spent hiding away in my tent, most had been lost to memory, though a few I recalled as glorious victories: the Jordanian clifftop; the Californian sea cave; the centre of a French *rond point*; the ramparts of a ruined Ottoman castle. Others I remembered as stonking defeats, and

I'd catalogued these as if they were dated horror movies: The Night of the Fire Ants (El Salvador); The Dawn of the Scorpion (under my Therm-a-Rest, in Argentina); and The Midnight of the Flood (Australia).

On this occasion, when the footfalls turned out to be a few, wandering kids, and *not* an axe-wielding sociopath, there was a sense of escape that washed away all the fear and seemed to make the whole process ecstatically worth it.

The road to Yangon offered up a man and a cow sharing a rickshaw. There were drunk soldiers ambling along, arm over arm, guns swinging from shoulders; beautiful flower sellers with uplifting smiles; and strings of monks in brick-red cowls claiming food. When I stopped for a bite too, a girl shot to my side, holding a faded, dog-eared pamphlet, entitled *English for Ladies and Gentlemen of Business*.

'Do you have any rubies or gems to trade?' she asked, reading from the book. I shook my head, took it from her and leafed through to find the appropriate response.

'I'm afraid, madam, the matter is quite one-sided.'

Inside the front cover, a passage read: 'These days the hill tribe people are far-seeing, they come down to the plains to visit, spreading markets, like us.'

There were fewer anachronisms left in Yangon, where modern technologies had been taken up fast as Myanmar opened to the world. Only a few years ago, the internet was virtually non-existent outside the capital, and mobile phone SIM cards cost $200. Now, I wandered past a clothing store called, curiously, Facebook Fashion, complete with the Facebook logo. An 'Epson' sign had been laid over one of the giant Buddha effigies inside the Shwedagon Pagoda. There was even an 'Apple Store', but I was happy to see that it had only espoused the name and dealt in fractured circuit boards and dusty radios.

I left Yangon a few days later with a little stomach trouble, an occupational hazard when you have the appetite of a touring cyclist, which is to say epic and indiscriminate. A fever and stomach cramps suggested that things were about to get explosively worse. I found a guesthouse in Okekan – all would be fine if my imminent coma was near a toilet. The owner looked at me over his spectacles and kicked at the dust.

'I have a room but I'm afraid you cannot stay. No foreigners.'

'Please! I'm sick and there's nowhere else to sleep,' adding a short stagger that foretold some medical disaster and a great deal of trouble on his doorstep.

'I'm sorry. The soldiers will punish me.'

Great, I thought, and cursed the military junta, adding my woes to their various sins. Forced land confiscations, torturing advocates of democracy, recruiting child soldiers, and now this.

It was a grim night, camped beneath apples trees, doubled over in pain, and I was exhausted when I returned to the road the next day. I wasn't the only one. Near Pyay, women were tugging up rice from flooded fields, more women were building the roads, and yet more were working in shops and, at the same time, caring for children. Men looked less burdened; they loafed, drunk, in the shade of plastic drapes advertising Grand Royal and High Class whiskey. The brand's respective slogans were 'Enjoy life!' and 'Taste of life!' Ironic, given how corpse-like these men had become.

*

The rules were being rejigged so frequently in Myanmar that it was hard to know which pockets of the country remained off limits. Chin State was one such enigma. Mostly cloud forest, tucked in beside India and Bangladesh, just north of Rakine State, Chin State hoists itself up, as if posturing alongside the Himalayas. Chin territory was a collection of chiefdoms before and after the British laid claim to Burma, and in the thousands of miles of greenery were just nine townships. For years, foreigners had been forbidden to enter Chin State without a guide.

I was aching for wildness, for the proper sounds of the world, but I still dithered at the junction. I had the legs; I just wasn't sure I had the time or the nerve for this. As usual, I was racing against a visa, and on my map, the roads through Chin State looked like the trace of a fatal cardiac arrhythmia, which could only mean heart-thumping climbs. They were unsealed too and the cloudbursts of the monsoon spelt mud. July was the worst month of the year to be here. And yet none of these pressures seemed reason enough to miss out.

I climbed into cloud. Bright and heavenly at first, then leaden and dreadful. Drizzle steeped my beard and made morning cobwebs of my arm hair. Briefly a breeze threw off the cloud and there was an instant of massive mountains, luscious green, furling into the distance. The stilted houses of Mindat flickered into view a bit later, their roofs electric-blue and green and brick-red. Villagers emerged. An elderly man was shouldering a rifle with a barrel at least half my height, a thing from antiquity that clashed again with the notion that change was in the air in Myanmar. Two women shuffled down the road grinning toothlessly, a swirling pattern of lines: their faces were tattooed, an old practice in the Chin hills, which helped prevent women being kidnapped by neighbouring tribes.

A man pointed me into his home and I accepted the invite. His house was wooden like the rest, tin-roofed and wind-rattled. I crouched by a popping fire, watched intensely by four children. In the adjoining room were fantastical photos of these kids, their faces pasted eerily onto the bodies of other children dressed in suits, on boats or at the seaside. The man was a pastor and there was a picture of a sad-eyed, lightly bearded Jesus on the wall too, next to another deity, a sad-eyed, baby-faced Avril Lavigne, whose image is, I promise, almost as common in remote villages around the world. Steam coiled off my clothes and blended with the woodsmoke, and we all silently wondered what I was doing here. Men and women arrived to shake my hand, chewing the enamel-dissolving betel nut. The Burmese have the easiest smiles, and the worst teeth.

'You are very good!' a man said, the village's prodigal son, back for a short stay from his home in the US. 'You leave your comfortable country and you come to ours and you live … like this!' He gestured at me, at my arms, muddy and peppered with ant bites, and my legs, greased in chain oil. A lady stoked the embers to keep me warm, and a girl ran off with my empty water bottles and brought them back brimming. Outside, a stranger was oiling my chain, another had sheltered my bike with a tarp as it began to drizzle.

I said goodbye to the pastor and put my warmer knees to use, climbing to Hakha, which was folded over a ridge like many of the villages here, perhaps because of the danger posed by landslides. An hour later, a flurry of fist-sized rocks came bouncing down the mountain up ahead. I looked up, chose my moment and pedalled madly, then turned to watch. Earth began to flow from the mountain and then at once a huge overhang of mud and tree roots subsided. Entire trees came crashing down. I turned on my heels and ran, pushing my bike, whirling around to see that the road was now lost entirely. I spent the next hours playing 'what if', recalling that day's small events, giving myself reasons to have passed the road moments earlier when the mountain would have claimed me within it.

The hills were sharp and relentless, the road now inches deep in mud. Leeches fell on me from above, then fell off, bloated with blood. At night I camped in the forest, wrenched awake every now and then by the rumble of a distant landslide, or the mutter of bouncing rocks. From spear and elephant grass I climbed into a forest of holly and rhododendron, then dwarf bamboo and pine, a nourishing habitat, that, paradoxically, can get so nourishing that it forces the region into famine every 48 years.

Mautam means 'bamboo death' in the local Mizo language. The bamboo here flowers around twice a century, bringing a sudden abundance of rats for whom bamboo seeds are delicious: an all-you-can-nibble buffet. The rats then flood the bamboo forest, overrunning crops and village grain stores. Once it was the stuff of folklore; an act of God, armies of rats sent in retribution. But people noticed that the famines were cyclical: 1862, 1911, 1958. Two years of preparation went into the last *mautam*. In just four months during 2009 there were region-wide campaigns to kill rats in the Ayeyarwady Delta. Villagers collected more than 2.6 million rodents, cashing in their tails for a small cash reward.

*

I arrived, permit in hand, to Tamu, the Burmese border post, on the last day allowed by my visa. All of India was just down the road. Inside the immigration office, there was a large, wooden plaque on the wall, entitled 'The Myanmar Spirit'.

> The simple-minded Myanmars have no envy against those having fair complexion
>
> Nor hatred for the brownishs; Nor differentiate with the blackishs; Nor hostile to those of different faiths.
>
> They have brethren love & affection and respect equally for all
>
> Irrespective of above all, if the affairs of our country, nation, land, history, culture, religion and preaching are interfered with a foxy trick to implicate national policies, it

> would be dealt with severely however great or small, black or white and so on with all the might but without a single word to finish to the end even if we are left by a single person with full of injuries lying in a pool of blood

I left Myanmar unclear who these 'Myanmars' were. Ancestry cannot be the answer, because where do you draw the line? In the broadest sense, we are all related. I thought about the slippery dream of nationalism. We've always been clannish, but nationalism is surprisingly modern, growing from the Romanticism of the eighteenth century when an idea emerged that there was a national soul, akin here to 'The Myanmar Spirit'. The spirit of a nation is tied to its art, literature and traditions, and to the mishmash of things we call culture. Nationalism helped mould identities and was crucial in spelling the end of colonialism and in reinforcing the right to self-governance. But it is, of course, a lie. Everywhere is, and always has been, multicultural.

It is easy to say that self-determination is *our* right, but who are *we*? The 'national policies' hadn't helped Paw Htoo as she lay dying in a foreign clinic on a hill, but then, as a member of an ethnic minority, forgotten, or even targeted, by the junta, perhaps she didn't belong to this Myanmar. She didn't meet the criteria, she disrupted cultural myths – the integrity of a nation, the truths and half-truths and fictions of its history.

In this sense, was Myanmar very different to Britain or America or Thailand, or any nation with a skewed sense of what it is, how it came to be, what it values and needs? Every nation tells a story, like every person does, and we self-mythologise for the same reason that nations do: to enhance our sense of uniqueness, and for cohesion. The slogans that nations use today to attract tourism and investment capital hint at what the cultural hegemony values, at the stories we want to tell.

Scotland: *A spirit of its own*. Bhutan: *Happiness is a place*. Turkey: *Be our guest*. Canada: *Keep exploring*. Egypt: *Where it all begins*. America has romanticised the frontier, which feeds the notion of a land of unlimited opportunity and delusions of great social mobility. The stereotype of the British in the Second World War, all pluck and decency, denies the fact that we had help, and quite a lot of it from colonies that we'd subjugated. Stories build nations and they strengthen a sense of identity, of which nationality, of course, is only a piece. Perhaps identity is a negotiation that never ends, where both insiders and strangers have a say, and perhaps it's restless because its context is, because the world is – like sand dunes, winds and human beings.

9

Hydration

'And can I ask … what is your good name, sir?'

The official at the Indian border post at Moreh stood up from behind his desk, smoothed out his moustache and thumbed my passport mindlessly. I told him my good name.

'And can I ask … what is your religion?'

'None,' I mumbled, feeling sure he wouldn't take this well.

'None?'

'None.'

'Not any religion? Nothing at all?'

'I'm not religious,' I said, dropping my head, inexplicably ashamed.

He returned my passport and, with a small 'that is all' wiggle of his head, I was shown the door. Outside, my bicycle had a new attachment. A group of children, their hands fidgeting over my seat post and grip shift, twirling my pedals, tugging possessively at my handlebars. I stepped into the yard to reclaim my bike with the peak of my cap pulled down low, too low, it turned out, to spot the top of the gate. There was a thunk, followed by eye-watering pain from the crown of my head. I dropped heavily to the ground. The guard was quickly by my side, plucking me to my feet.

'You see, good sir?' he said, concerned. 'Everybody need a religion.'

*

If any would do, maybe I'd pray to the sun gods. The monsoon was grinding me down; my bike too. Each day brought another click and jangle, rust was flourishing, several teeth on my chain wheels were chipped and my chain had worn saggy. After a pedal fell off, I'd replaced them both with the kind used on cycle rickshaws: they'd cost 15 rupees a pop (around 18 pence – rust already included) and each weighed more than a kilogram. It was like cycling with *A Suitable Boy* nailed to the soles of my shoes.

In Moreh, men tugged wooden carts loaded with grain in burlap sacks and rice and vegetables, beetling through puddles, backs brilliant with sweat. The fruity warble of Bollywood films blared from eating houses until the sudden silence of another power outage. As a rule of thumb, the frequency of blackouts in Indian towns is proportional to the number of tangled, electric cables hanging over the streets. Moreh was jumbled by cables and stricken by blackouts.

In India eight states herniate from its north-east and only a 14-mile-wide umbilical cord of land, the Siliguri Corridor (also: the 'chicken neck') couples this part of the country to the rest. To many, it is tenuously Indian in other ways too and the reputation of the north-east has been dominated less by its cultures and ecology, than by its politics. Its internal and external borders have been contested and the whole region is often viewed simultaneously as a neglected and defiant place, where the Indian army have been known to carry out extra-judicial killings, and where insurgents still obsess over questions of nationhood and identity.

My own identity was on the mind of the pot-bellied owner of my hotel.

'So you are a roamer!' he proclaimed, hands on hips. 'And is there a wife for you, back in England?'

I shook my head.

'Ah ha! A bachelor!' he said, looking dangerously close to winking. I agreed. It was better than 'single', with its rather lonesome connotations; in India, I was yesteryear's 'player'.

'You're going to Imphal?' he asked.

'I am indeed. How's the road?'

'Oh fine, fine. Nice and flat for you!'

'Flat?'

The hairpins on my map insinuated otherwise.

'Yes, yes. Totally flat. But bring some warm clothes, sir! It's very cold, when you're up in the clouds.'

*

Moreh could be as entertaining, and inscrutable, as its residents. Hotels advertised 'lodging and fooding' and the idea that *food* can become a present participle is a nice one, I think, because it suggests that it could also become a verb. As in, 'Man, I'm starving. I'm gonna food the hell out of this place!' Or, 'Brian, stop fooding! You'll make yourself sick!'

In the street outside my guesthouse, a poster advertised the services of a sexologist, who boasted cures for various carnal afflictions, from sexually transmitted infections to impotence, but it was the ailment of 'sexual devility' that tickled me most. For some reason, it brought to mind a burly man in a crotchless red devil outfit with a lusty glint in his eye.

Everything seemed to move differently in India, from the tuk-tuks nosing into stalemates at junctions, to the cows loping up and down the street with the calm of invigilators. My own pace was hampered by a sequence of police and army checkpoints, far more than I had encountered in Myanmar. Smuggling opium was rife here, addiction too, and Manipur had the highest rate of HIV in India.

'Hello, sir. You'll have some tea?'

I stopped beside the soldier, ahead of the fifth roadblock that morning. I considered his invite for a second. My Indian visa had been ticking since Bangkok and if I didn't get to Nepal soon I was going to run out of time, and then I'd either get kicked out or an official was going to get a sudden and discreet pay rise.

'Thank you. But I've got a long way to ride today.'

He thought about this for second. 'No. You'll have tea. Please, sit down over here.'

'I'm sorry ...', but he'd already turned to bark in Hindi at another soldier. A rifle was dangling from his shoulder and now the muzzle came to rest on my left thigh. An accidental shot, I noticed, would take out my knee cap.

He turned to face me again, oblivious. 'So you'll have tea?'

'Absolutely.'

'Excellent. You are a fine man. Come. Meet the captain.'

There could be no mix-up – aviator sunglasses, a galaxy of obsequious subordinates zipping around him, the captain leant back in his chair and smiled with the self-assurance of someone rarely messed with. With a faintly bemused look, he studied one of his soldiers questioning me, tapping his pen archly on the desk as I fended: 'What weapons do you carry?' 'Father's profession?' and 'Vehicle registration?', which might have ended in a tense stand-off had I not scribbled FKNOWS into a box on a form.

The captain pointed to the tea pot on the desk and a soldier sprinted off to brew more.

'Welcome to India! You beat us at cricket, you scoundrels!'

'Did we! Sorry! Don't worry, it won't happen often.'

'Ha! You have exciting life in England, no? This is simple life for us in India. In England you go to casinos all the time, you run about doing exciting things.'

'Honestly, I don't go to casinos.'

'What do you think of this place, it's pathetic, no?'

'Well no, it's—'

'It's pathetic. Most pathetic place in whole of India! Do not trust these people, do not even trust the police! People are so backward here. It is like your country in the time of Charlie Chaplin.'

*

As I arrived into the town of Pallel, there was a snap. I looked down: my chain had done the snapping. It was growing dark as I crouched beside the road, trying to work my chain-fixing tool, which had been bent out of shape and no longer appeared to fix chains. A young man drew up on his motorbike and sighed.

'Too late to fix this. You want to stay with me? Come! My family having fish curry!'

Lightson was nineteen years old, a student, five foot two of bookish charm. We sat on the porch of his house, in a tiny gated courtyard, as a numerous family of uncles, siblings and cousins squatted around us, a circle of steadfast, nut-brown eyes. I asked Lightson what he felt like doing after his studies. He looked hesitant.

'In India, you pay to *get* a job. This is crazy! Everything's about money. Especially here in north-east. So many outfits. These people are terrorists, if you have a business and you don't pay them, you suffer. I don't know what to do.'

'Why not move? India's a big country, maybe you'd find opportunities elsewhere?'

'I want to get out of here, I want to travel. But I can't afford to.'

Lightson gazed at the tamarind trees beyond the fence.

'I am like a frog in a well.'

That evening, the men sat in a circle, cross-legged, in a small outbuilding, while somewhere in the back pots breathed steam and the shapes of women moved to and fro, like a shadow play. We scooped up rice and curry with our hands, ate fast and wordlessly.

Lightson's aunt was an old stooped woman with something quick and shrewd in her eyes. She lurked behind our backs, ladle in one hand, mighty pan of rice in the other. At unpredictable moments she would snap forward to tip another heap of rice onto a plate, with a velocity that belied her years. With bellies stuffed, the men whipped their hands in front of them to bar the ladle's path, but she was tenacious, hovering in the shadows and coming for me most often. I imitated the others but she didn't play fair, switching sides at the last moment or batting my hand away. Mound after mound of rice arrived and soon, I was fooded to the max. In the half-light, I sensed that the small Indian woman was grinning.

I fixed my chain the next day, hammering links together, before joining the road towards Imphal, the chief city in the state of Manipur. Minibuses and trucks drew level, their drivers gazing at me to the neglect of everything else. You must be bullish on India's roads, but you must hope that your bullishness is slightly more or less than that of the driver approaching you at speed; equal bullishness can make a horrible mess. One motorcyclist had all the bearing of a shark in a shoal of mackerel. He stared and stared and stared, unfazed as cars scattered, blaring horns. I made eyes at him, like he was crazy. Eventually, he called out 'Hello mister! What is your hobby?'

For the first time in my life, 'survival' now fell under that rubric.

The road signs were as ominous as the dints and scrapes on the rickshaws and cars. 'Drive, don't fly', they began, then the tenuous 'If married, divorce speed' and 'Speed is like a knife that cuts life'. Boozing was condemned through comedy and double entendre: 'Drive horsepower, not rum-power' and 'drunk drivers are bloody idiots'. The motorcyclists ignored the most vital of these signs: 'If everything comes towards you, you are in the wrong lane'. There was little sense of right or wrong, however: the lanes were for pedants and the horns were mind-breaking. One of the first things people did in India on buying a new car was to remove the horn and install a new one, trebling up on decibels. I would never adjust to the horns in India. I tranquillised myself instead, donning sunglasses, with a Buff about my face and headphones in my ears. Nina Simone was the salve to my tinnitus; she poured soulfully into my ears.

I unplugged near Imphal as a young man cruised beside me on a wine-coloured Royal Enfield motorbike, printed with the Union Jack.

'You want red-light place? Girls and she-males!' he shouted over his shoulder. 'I hook you up, no problem – 3,000 rupees!'

'For a she-male or a woman?' I asked.

'Maybe we find one of each!'

I settled on exploring Imphal instead. Litter stank in heaps, fat rats surfaced in blinks of street light. Buildings looked as if they were frozen in a moment of collapse, each a disorder of lurching bamboo scaffolding, sheet metal and tangled cables. The city had long been given to gusts of discontent and the TV news in my guesthouse lobby played scenes of stern-looking men marching double file in the road protesting what they said was an incursion of outsiders from other Indian states.

I followed the road into Assam, then a new, wild-sounding and even damper Indian state, Meghalaya. Signs pleaded 'Educate girl-child'. The forest receded when Silchar began. I stopped to ask a man for directions to a guesthouse, but he dropped his gaze and said, 'I'm sorry. I am not a civilised man.'

In Silchar, men chewed paan in molecule-like gatherings around any kind of nucleus: a conversation, an argument, a rickshaw crash, a card game, me. Quickly, it was established: I'm British, I can't speak Hindi, I'm going to Darjeeling, yes I'm a bachelor, yes I'm alone, yes really, yes completely alone. In most places, the mere fact of my bicycle was enough to cause exclamations of surprise, but in India, surrounded by this tightknit audience, cycling was less startling a detail than the fact that I did so alone.

I was reminded again of the distinct concept of privacy here when I caught a snippet of a conversation in the street.

'I'm off to visit Anna.'

'Who's that?'

'You don't know her. She's in hospital.'

'What's wrong with her?'

'Something with her ears, I think.'

'What's wrong with her ears?'

In India's bustle, it didn't seem unreasonable to know about Anna's ears, even if you weren't sure who Anna was. On the streets, my bike horn was often playfully honked by passers-by. My journal was filling up with names and addresses of men who insisted it was in my interests to know. I'd heard that modern Indian languages often do not have a specific word that precisely captures the meaning of privacy; they're usually some variation of the words for isolation, intimacy or secrecy. This might feel frustrating for the outsider, especially if you hail from a country with an apparently more definite model of private life, like Britain, which fanatically mines online personal data, is stuffed with CCTV, invests in facial recognition and commodifies biometrics, and which has politicians asking the same question an Indian might: 'But what do you have to hide?'

*

In the hills beyond Silchar, I arrived at a roadblock. The afternoon rain had passed on, leaving a frail, ashen light. I was high now and I felt the cold of the altitude in my bones. A policeman was all questions, then leaned forward and laid a hand on my shoulder: 'We salute you and your amazing adventure! You are like Braveheart, in the Mel Gibson movie.'

Behind a few wooden huts where the policemen slept, the forest fell and then flattened into sun-kissed lakes and rivers, darkening here and there with stands of sal and teak. 'Bangladesh!' The policeman beamed. Of course, but with a single-entry visa for India, it was out of bounds for me.

'You need a place to sleep, no? We are honoured you chose us. When my grandchildren see you on Discovery Channel, I will say to them: "That man stayed with us!" You are a legendary man.'

I protested, but he stopped me with a raised hand. 'Legendary,' he assured me. 'And now, you must give me selfie.'

We posed together for the camera. On the screen a dicey, dead-eyed man appeared beside the policeman. My face was flaking and ruddy. My hair was the crazy halo of an electric-shock victim. My expression was a strange blend of emotions: there was shame, melancholy and, unfittingly, lust. I looked like the kind of man who might present at 4 a.m. to an emergency department with something embarrassing in his rectum, perhaps a root vegetable.*

---

* Every doctor has an object in orifice story that they wheel out for dinner parties without contravening the General Medical Council guidance on good medical practice article 50 of Domain 3. Mine is a rectum and a certain root vegetable that may or may not be a butternut squash. At this stage in the party I'm assailed by a great gush of questions concerning the logistics of such an anal supplement. I'll just leave it to your imagination.

The policeman led me to a small shack a little further down the road. He ducked inside, chickens fled raucously, and he walked back out gripping one about its neck, frowning through a casual execution. Then he brought me a lit candle, which illuminated cobwebs and a rug of chicken shit. But it was roofed! And all mine. And from the window-hole (later, my en suite) the lakes of Bangladesh shimmered like flakes of mica. My groundsheet screened the chicken shit and I burrowed into my sleeping bag and read *Desert Solitaire* by candlelight as rain applauded on the sheet-metal roof.

'Today, no cars,' said the policeman the following morning, strutting about the silent run of mud as if to verify the point. This was a protest, a 48-hour shutdown on Indian Independence Day called by the Hynniewtrep National Liberation Council (HNLC), a local militant organisation reputed to carry out their own justice, padlocking men through their ears and ordering them to stand in town centres for a week at a time.

Back on the road, the horns were missing, but not missed. Nobody risked their future to ask about my hobbies. The sky had turned the colour of shark skin, a great depression was stirring in the Bay of Bengal – a belter, I was told, even by monsoon standards.

The traffic began to move after a couple of days, but then it didn't. A jam of trucks, painted the orange, green and white of the national flag. Christian crosses were propped in windows and small signs reminded us to 'use your horn' and that 'India is great'. Up ahead a waterfall frothed through a dun-coloured gash in the forest, breaking rockfalls powerfully about it and flooding the road.

'This road closed for two days now,' said a trucker cloaked against the spitting cold in a chequered red shawl. 'Last year, we were stuck for a week,' he said, as if admiring the authority of Indian skies. Boulders big enough to crush three men lay scattered over the road, and some had tumbled further, smashing down trees and baring streaks of earth. It was compulsive viewing, like the coin-pushing machines of seaside resorts: as each rock falls you guess which others will be sent tumbling too, with jackpots of rock sometimes pelting down the mountain. Three hours later, a digger arrived, scooped up muck and reopened the road.

I rested for a day in Shillong, the capital of Meghalaya, a high-ish town I was quite fond of, garnished by a very blue cathedral, the odd café and even a pedestrianised area – a concept drivers dispensed with – and a no-beep zone, which was celebrated through the medium of beeping. My thoughts turned to the rains. The monsoon was getting wetter. It was something to suffer or submit to, not something I could reasonably go round or wait out. And if I *had* to experience weather, maybe it was better to explore just how extreme the monsoon could be. With this in mind, I set off for Cherrapunjee, a small village south-west of Shillong where there had been no sunset to speak of for months.

Meteorologists can't agree on the exact whereabouts of the wettest place on earth. Parts of Hawaii get pretty drenched, but Cherrapunjee is a front runner too, and its neighbour, Mawsynram, is sometimes given the crown. The annual rainfall of Cherrapunjee is recorded in millimetres, which seemed cute, as if an extra two or three would make much difference. On an average year, around twelve *metres* of rain fell here. Welsh missionaries set up base in Cherrapunjee 175 years ago, but they soon gathered up their dank clothes and mouldy bibles and retreated to Shillong. I'll repeat that: the *Welsh* fucked off because it was too wet.

As I neared the village, water slicked every rock face, drenching ferns and stripping topsoil. I was flooded too; rain had invaded my raincoat through the seams. I heard only water for hours, a ceaseless puttering in my ears. For a few seconds I glimpsed the high rim of limestone that marked the edge of the Sohra plateau, a battlement of rock striped white by waterfalls, but then nimbostratus closed in again, and there were strobes of sheet lightning.

Stories brewed here – thundery parables, keeping the local Khasi kids on their toes. A woman dressed in white was said to fade from the rain mists: the spirit of the forest. They say a man once pissed in the forest, desecrating her realm. He went slowly mad and spent his last days cooking meals for a family he didn't have.

The Khasi hills loom fast out of the surrounding plains, and wind and clouds are thrust upwards, scaring the shit out of any pilots in the vicinity. To the south, there is a funnel-shaped catchment, which helps explain why, in just one month, July 1861, nine metres of rain fell here. That year's total was a world record of over 26 metres. By contrast, the wettest part of the UK, Snowdonia in Wales, averages three.

The clouds finally parted as I moved north-west, into the 'chicken neck', and the body of India. The road ahead turned white in the sunlight, or appeared to after months of monsoon greys. The shadow cyclist appeared beside me again and then the floodplain of the Brahmaputra River began. One of Asia's greatest waterways, he* drifts down from the Angsi glacier through Tibet before fattening up through India and spilling into the Bay of Bengal. The annual floods are expansive and I cycled through a water-world where just the road, forested islands and the odd village were spared. An elevated railway had become the only thoroughfare between villages, and hundreds of people, carrying bundles and firewood on their heads, marched silhouetted along its course. As dusk set in, I camped

* Unusually for Indian rivers, the Brahmaputra is denoted male.

on the edge of a forested island; a few kids sat watching me thread poles into my tent. When they left, a green whirligig of fire flies took their place.

The next day land rose from the water and the Buxa forest gathered, the trees trembling unaccountably until I spotted monkeys playing high jinks in the boughs. A sign reminded drivers that elephants had the right of way – a useful tip considering the comparable obstinacy of Indian drivers and elephants – and then three boys on a motorbike drew up and propositioned me for a selfie. When the job was done, one kissed my hand, the other grabbed it and held it to his heart.

The road swung close to Bhutan now and I spotted a smaller road forking off to breach the country. I didn't have the coveted Bhutan visa, which came with a price tag of $250 a day, more than I'd spend in a month, but perhaps I could sneak in, nose around, export some of their fabled national happiness, import some of my own. My lazy prejudices about the Buddhist Bhutanese – that they were a gentle, peace-loving mountain people – suggested I could try and that they wouldn't shoot me for doing so.

Getting past the Indian post was easy, and for once I celebrated a rackety Indian road – I was well concealed behind shambling goats and streaming rickshaws. Beyond, a big arch lifted, etched with dragons. A gasping, battle-worn truck bumped past and I took my chance, pedalling hard to keep up with my cover, slipping beneath the arch, unseen.

I climbed for a few miles and looked out over the stone border wall that stretched across meadows and lapsed into a cold-looking river. A small road was signposted 'Kanyo Thang' – and I thought about how often small towns around the world sound like they could be members of Wu-Tang Clan. (cf. Keetmanshoop, Namibia). I followed it, pausing by a class of schoolchildren dressed in *gho* robes, all huddled about a teacher who was reading to them in the sun. The world's politest children filed past, chorusing 'Happy journey, sir!'

But to journey into the heart of Bhutan would risk getting caught and deported, and I was a bit worried that my unhappiness about such an outcome may not be legal in Bhutan, so I trundled back the way I'd come in – poof! – transforming from illegal to traveller as easily as I'd managed the reverse.

*

There is no easy way to approach Darjeeling by bicycle – all roads are banefully steep – but my heart was set on the roughest climb, first through the Neora Valley National Park, then down to the Teesta River and up again from Kalimpong. From the tiny village of Matelli, I took a trail to Gorubathan. A few women, hip-deep in tea, glanced at me sideways, picking like automatons. The track ended in a footpath and I wheeled my bike to a cliff with a thin path running across its face. Two men arrived behind me, and without conference they grabbed parts of my bike and the three of us hefted it along. My foot lost purchase, there was a panicked, somehow successful scramble, and my bike was saved at the last second from tumbling into the river below. At the road beyond, the men branched off without waiting for a thank you. 'The forest people will look after you,' one shouted, over his shoulder. And then, outlandishly, 'Watch out. Tiger! Elephant!'

In a village halfway up the next valley, a few men offered me a hut for the night. I'd be sharing with an elderly man and whatever the two suspiciously rat-shaped things were, scarpering up the walls when I opened the door. They were enormous, toe-amputating fuckers. I'd have been better off in my tent but it felt rude to shun their hospitality. The price for such good intentions was a fresh slew of mosquito bites and a night of tense, wakeful moments, full of old-man snores and scuttling shadows.

The valley kept on the next day, sharp and pine-flanked, the road making ever tighter switchbacks. At Lava, a Buddhist monastery was perched on a hillside and young monks waved me on as I mulishly climbed over the village and then careered downhill, glimpsing wild peacocks, flustering through the trees.

And then it began: a climb of 1,500 vertical metres in just 13.5 km of tarmac, an average gradient of over 11 per cent, corners nudging 25 per cent. By comparison – and to make me sound plucky – I should mention that the Col d'Izoard and the Col du Tourmalet, two of the Tour de France's most defeating climbs, average a piddling 7.4 per cent and have tested riders carrying fewer pots and pans than me. It helped to nurture a sort of wrath at the road, at the mountains, at myself, which recalled Jens Voigt's rebuke 'Shut up, legs!' When a car overtook with a sign in its rear window that said 'follow your dreams' I thought exactly this: fuck, piss, wank, balls.

When the swirls of cloud are gauzy enough, Darjeeling is one of the most dramatic big towns in the world. Forget the idiom, of *hill*-stations and foot*hills* – British mountains don't rise to half its altitude. The day after I arrived, a breeze kicked out the last of the cloud, showing a deep valley and threads of roads. Above the town lifted some of the tallest peaks on earth. Kangchenjunga is particularly mighty, India's highest mountain and the world's third, making even the prodigious peaks of Sikkim to the north look puny. It's a sight branded by the travel writer Jan Morris as 'one of the noblest experiences of travel', and one that 'has moved generations of pilgrims to mysticism, and even more to over-writing'. So I'll leave it at that.

I mooched around the zoo for a while, but the animals were sulky, mangy and traumatised, especially the jackal, which was pacing out a life sentence. There were white people here going full Indian in glittering saris. I wondered how long the camouflage and affectations would last, or if they'd be drinking in British pubs in a few weeks' time, still jangling their bangles as they sipped on Carling and watched the Premier League.

I'd been in touch with a pair of Americans, Mike and Chris, who had also cycled from Myanmar into India. Since Myanmar was only just opening up, it was a journey few bikers, besides us, had made. Feeling lonely, and bored of the box of annoying thoughts atop my neck, I was looking forward to having company, and we were soon sitting on a balcony of a guesthouse above the lights of Darjeeling, whiskey in hand. They were suffering the monsoon too. 'Bikes rusted. Clothes mouldy. Scrotum full of fungus,' moaned Chris. Mike nodded sagely, then frowned as he looked down at his own crotch.

We spent the evening as touring cyclists do when left unsupervised: spinning yarns and war stories, our own and second-hand, but my Indian visa was nearly up so I left for Nepal the next day. Impatient for Kathmandu, I stuck to Terai, a flat grassland, mostly, fretted by rivers and mottled on my map by occasional forests and swamps. After the monsoon rains, white-topped kans grass had grown fast in quivering swathes. Women crouched, weaving plant fibres by the road, and kids raced me in the evenings, gangs of them on rattling bicycles many sizes too big. I let them win, occasionally.

Terai extends from Nepal south into India and for centuries it was ravaged by disease – a malaria hotspot – until the arrival of the insecticide DDT in the mid 1950s. Then, Nepalis from the mountains, Tibetan refugees, Bangladeshis and other Indians moved in with the indigenous Tharu and Dhimal people. The influx created tension. I'd noticed before how climate and terrain can have a bearing on disease, but here was the converse: disease can shape a place too.

One afternoon, I spotted a tiny child with a backpack standing by the road to my left. A school bus was grumbling at a standstill on the other side of the road and, with a sense of doom, I saw that he was about to run towards it. When he bolted there was no time to sound the bell, even to shout. I was steaming, a tailwind helping me to 30 kilometres an hour. I turned hard to the left, panniers lurched, and there was a swish as his school jacket skimmed off a pannier. And then he was stepping onto the bus, unaware that he'd been so close to earning the nickname 'tready' from boys in his class by virtue of his face being permanently tread-marked. No, it would have been worse than that. Little 'tready' wouldn't have made it.

*

Kathmandu: you'd be hard-pressed to build a capital elsewhere in Nepal, as the city collects in one of the largest pockets of mountains available. I surveyed the dusty, smoggy lay of it from the hills, glad at the prospect of some time off, especially when I realised the date, 13 September. I'd forgotten my birthday for the second year in a row.

But Kathmandu would have to wait – first, I planned to visit a hospital ten miles to the south. My friend Ian's grandfather had founded Anandaban, a hospital for patients with leprosy, in 1957.* When I arrived, the air was perfumed by pine trees, which stood tall around the lab and wards like benevolent minders. Birdsong rang out. 'Anandaban' means 'forest of joy'. The view through the pines glinted with the tin roofs

* Let's get this out of the way. Leper. It doesn't sound great, does it? The preferred term today is Patients Affected by Leprosy, something of an acronymic master stroke to my mind, considering their ostracisation. Yesterday's lepers are today's PALs.

of houses, making a firmament of the hills. A forest fire had almost torn through the hospital a few years ago. The patients, however, had prayed, and in this way, according to a pamphlet I'd found, disaster was averted. They say a leopard was seen dashing from the flames.

In the reception, photos of visiting dignitaries adorned the wall. One of Princess Diana greeting patients caught my eye; it was three years before her own death. Dressed in pink, she crouched at the feet of a man, reached out, and touched his diseased hand.

In the women's ward, Parbati stood out because she was young, with a gold nose ring. Her face, lightly freckled, brightened with a smile.

'Namaste.'

When Parbati bowed, the remains of her hands rose up to her face, palms pressed neatly together, finger stumps pointing at the ceiling. Leprosy had ravished her limbs: her wrists were forced into flexion and her hands would have been claw-shaped had she not lost her fingers years ago. Flesh had died piecemeal since childhood. She was detained in bed because she had no toes left either.

Parbati had been living with her father and brother in a remote mountain village when she first noticed her skin change. Villagers noticed it too, and whispered that the deformed girl deserved it: her ancestors had sinned and this plague was her dues. They'd scorned her father and brother for allowing her to remain in their home. She is bad luck, they'd said. Her curse could spread. Parbati became a recluse.

One day, her brother brought her a battery-powered radio, something to keep her diverted while he was out working. She crouched down in her world of the hut, ear to speaker, straining through static like frying fat. A special health announcement caught her attention – the symptoms of a disease: numbness, weakness, patches of pale skin. She recognised the symptoms as her own. Apparently, there was a cure.

A few days later, she convinced her brother to heft her into a basket strapped to his back and begin the seven-hour trudge through forest, over rivers of snowmelt, to the nearest road. After three days, she reached Anandaban and started the treatment that would have saved her hands and feet a decade ago: antibiotics.

Things were looking up for Parbati – she was being treated with a combination of drugs and had been trained by staff how to knit without fingers, balancing the needle between her stumps, artfully waggling it in any manner she pleased. The hat she was finishing had four colours of wool and a bobble on top. Parbati wanted to open a shop. As I admired her work, she fished out a plastic bag with a dozen finished hats and asked if I wanted to buy one. I did.

Leprosy is perhaps the most mythologised of all diseases. It has been called a curse, wildly contagious, untreatable and inexorably fatal, yet it is none of those things. Less than 5 per cent of people are even susceptible to catching it and many sufferers cannot transmit the infection. It's been singled out since Leviticus (13:46): 'All the days wherein the plague is in him, he shall be unclean, he is unclean, he shall dwell alone, without the camp shall his dwelling be.' Leprosy has long been considered a disease not simply of the body, but of the soul.

In 1873, a 32-year-old Norwegian upstart called Dr Hansen established that lepers were not victims of heredity or the wrath of God. At the time of his discovery, he was the assistant to the award-winning Dr Danielsson, then the world's established authority on leprosy. Dr Danielsson contended that any idea of contagiousness was just peasant superstition and he even inoculated himself on several occasions using samples from diseased tissues. Because he never became ill, he proclaimed that he'd proved his point. But using histopathology techniques, novel at the time, the junior Dr Hansen discovered rod-like bodies within cells of leprous skin nodules. It was *Mycobacterium leprae*, the real cause of the disease.

That leprosy today remained so stigmatised and misunderstood rattled me, but Parbati's story seemed to champion a very human counterstrike too. There was Parbati's own resilience, a determination not to concede to a 'curse'. There was the devotion of her family. There were strangers in labs, bringing antibiotics to the world. But ultimately, it was not just antibiotics that saved her. It was a radio. It was a human voice hundreds of miles away, an echo in the transnational drift of information, the forebear of telemedicine and e-health that could revolutionise how we help one another. In simple terms, it was someone reaching out and someone else connecting.

*

Kathmandu, at last. One of my first tasks was to score a visa for Pakistan. In the Pakistani embassy I was told that this would be '99 per cent impossible' by a distracted lady who didn't expand on the 1 per cent opportunity. Eventually, I was given forms and told to send my application to the UK. I spent days forging statements so I could prove a bank account flush with savings. I faked hotel reservations, and concocted tour guides and tour agencies. I posted off my passport and the whole shebang of forms, reservations and 'evidence', but I wasn't reassured by the name of the company to which the Pakistani embassy had outsourced the visa application process and who were to keep safe these important documents. 'Gerry's visa dropbox' just didn't inspire much confidence.

Three weeks later I was told that some trifling form was missing – they couldn't process my application. With Tibet technically closed to independent travellers, my line home had been suddenly severed. It was hard to see how to get to Europe now, unless I flew somewhere. This was a setback, but hardly an impasse or injustice. For many Pakistani or Nepali citizens getting a British visa would be a process of wild complexity by comparison. Not far from my guesthouse, there was an office where Nepalis could, in theory, get travel documents – a whirlpool of people praying for one of the most useless passports in the world. The most irritating thing about my situation, for me personally, was that if my own country hadn't partitioned India, my travel plans would have been much easier to arrange. Does the cruelness of partition never end?

At the time I was staying near Thamel, that spirited enclave of Kathmandu. It was peak tourist season and Thamel was all colour – hikers in loud down jackets, decorated women in lurid saris, the scattered flames of sadhus, backpackers in migraine-inducing pantaloons. Roadside stalls glinted with singing bowls, curved Gurkha knives and mock-jade effigies of Buddha. As I moseyed along, pondering all the ways I could get home, I began to look more closely at Thamel, and my eyes were drawn, finally, to the shadows. Street kids. By day, they loafed about or played tag, and in the cold evenings, they sat, packed tightly together, like passengers in a Kathmandu minibus, planting white bags over their mouths, huffing in and out. The bags contained glue. I watched a few kids get woozily to their feet, stumble off, dim and willowy figures in the streetlight, like a vision of extra-terrestrials.

The next day I booked a flight to another destination where my passport would be warmly greeted: Hong Kong. It appeared that I could cycle home from there, and so I drew a new line across the world – this one reached up through China, pierced the 'stans, the Caucasus and finally Europe. But, afraid I wouldn't be able to pedal the breadth of China within the time a visa allowed, my eyes were drawn further north, to Mongolia. Why not?

Well, there was a substantial reason why not. I'd be hitting Mongolia for winter, a nippy spell, the nights getting down to minus 40 degrees Celsius in a good year. A bad one had a special and disturbing epithet: *dzud*, or The White Death, like the winter of 2009/2010 when thousands of the country's yaks ended the season quiet, grey and significantly harder than they used to be. By the spring thaw, a fifth of all livestock had turned to dead-stock in nights of minus 50.

So that's how I spent the next day in Kathmandu: in the trekking shops, searching for a down jacket that would keep me warm, or at least flatter my frostbite.

10

Memento Mori

It was the Hindu festival of Dashain and Kathmandu transformed, subtly at first. Residents were leaving for villages in the mountains and there were fewer people in the streets outside my guesthouse now. The stray dogs remained, and someone had given one of them a bindi. From rooftops, kids flew crisp-packet kites; they swished and swooped and glinted in the sun.

I was in a freewheeling mood as well. The sky was too blue to be fastened to a city: the Himalayas were calling. First, I dropped by the Mango Tree eco-camp, crossing a slatted wooden bridge in the terraced hills above Besisahar, north-east of Kathmandu. Mike, the American cyclist I'd met in Darjeeling, had been volunteering here for the last few weeks and, without time or permits to stray too far, we planned on riding the Annapurna Circuit together. I wanted wilderness, and this was a compromise: we'd ride beneath giant mountains, 7,000 and 8,000 metres tall, and climb Thorong La pass, 200 metres higher than anything I'd ridden in the Andes. But it was all very user-friendly, homestays abounded and the trail would be busy with hikers. There was even a road under construction now too.

'It's pretty easy. You guys will ride down and then it's flat,' reported the young girl perched on the steps of the village police post, sounding as assured as the adults inside.

'Flat?' I wondered how that was possible. Mountains blotted out half the sky.

'Well, you know, "Nepali flat": Up down, up down, up down.'

Several hours later, Mike and I were staggering under the weight of that truism, shouldering our bikes as we trudged up and down steps carved into rock. We'd missed a bridge and we were meant to be riding on the road that we could see quite clearly over the water, a promised land conspicuously free of steps, trickling with whistling Scandinavians who were having a much better time.

We gave up in the late afternoon, and found a room in a small wooden house circled by a garden of squash, sago and marijuana. The next morning, the mountains were honey-tipped as we returned to the grunt work, and more steps. A small breeze blew in and the rice paddies whispered that we were shitheads.

Finally, a bridge reunited us with the road and now we passed small waterfalls and their shimmering rainbows, and purple-tinged fields, stubbled with stalks of buckwheat. For lunch, we stopped at a teahouse to chow down the local dish, dal baht, steamed rice and lentil soup, the most flatulence-provoking food known to humankind and a terrible choice for people walking single file behind one another for days.

As we ate, a small boy picked up Mike's camera and then handed it to me. One of us dropped it, I'm not sure who, but Nepali kids are all thumbs. Ever met anyone who practises Vipassana meditation? I can tell you that it's pretty hard to piss them off. Even if you scratch, dent and jam their DSLR camera lens onto the body. Mike's breathing slowed, his eyes glazed over and a tsunami of peace washed over him. 'It's okay,' he said at last, and the dude fucking meant it.

The valley was steeper now, and pouring over with sunburnt fern, snarled like scrap metal. Crows cawed and the wind shrilled in the pines. We stopped at a wooden bench of artefacts for sale: two great yak heads with light bulbs for eyes;

pottery dulled by indeterminate decades; goat horns; a black necklace fashioned from the vertebrae of a snake. I claimed a baby yak's skull and cable-tied it to the underside of my handlebars. Something prankish in me wanted to make children cry and old women bring forth prayers. I was coveting Mike's bike too, a more avant-garde effort than mine. Mike had tied halves of a painted coconut shell to the cross tube, and there were vivid triangles of Buddhist prayer flags over the frame that snapped to and fro on the winding Himalayan downhills. A bamboo root from Vietnam, carved into the face of an old man with wavy hair, stared mystically from the rear rack. Nepalis sometimes pointed to this face, asking 'God?'

I straightened my skull so that children would have to look down the eye sockets. 'What do you think?' I asked.

'Dunno.'

I looked up. Mike was scouring the valley and hadn't noticed my accessories at all. I followed his gaze. It was almost dusk now and a fleet of grey clouds had thickened at the tail of the valley.

'Probably nothing.'

Seemed a decent guess: October was peak dry season and we'd passed a number of checkpoints and sunny Nepali officials on the way. There'd been no mention of storms.

Manang was near capacity and only one guesthouse had room to accommodate two bikers and the haunted skull of a baby yak. Trekkers shuffled up and down the one street, trading books and dropping into the village mini-cinema to watch films that seemed to be entirely about death in the mountains. That night I woke, needing to pee. I skirted the outbuilding towards the long drop, stopping on the way to peer up, hoping to trace constellations. The night was speckled, but not with stars. Snow was falling. Not a flurry, just a few pioneering flecks, floating on the wind.

I woke to see an old lady shovelling snow off our neighbour's roof. Mike jumped out of bed and began cursing far more than was appropriate for someone on the Buddhist spectrum. Locals and tourists alike trudged wonderstruck through the village, two thousand metres below Thorong La pass, where the trail dithered through the bluest bit of my map. Maybe they were thinking what I was: *If there's three feet of snow here …*

Goats descended to rootle for vegetation, and hikers continued to trot in too. With nobody leaving, the village was soon clogged with people, and with the power out, there was nothing to do but sit by a fire fuelled by yak dung, read, and drink butter tea. The following day the sky was back to its pacifying blue, and Annapurna II and III were panes of white. The peaks were smouldering now, snow lashed off by wind in feathery arcs. In another homestay, I found a television showing the BBC news: images of helicopters, aerial shots of the mountains about us. The voiceover: 'One hundred trekkers are missing in the Himalayas after deadly blizzards struck the region. A rescue effort is now underway. Trekkers from Poland, Israel and Canada, as well as Nepalese, are among at least 30 killed …'

With a major disaster unfolding on the trail, and nobody better informed than BBC journalists on the other side of the planet, trekkers huddled around Manang's few computers, eager for weather updates, emailing loved ones. Beside me, a girl blinked at a website displaying a list of names, each hiker coupled to their fortune … safe, deceased, unknown, safe, unknown, deceased, unknown …

Helicopters began buzzing through the Marsyangdi valley, minutes apart, olive Russian-made choppers, owned by the Nepali government, along with red search-and-rescue aircraft too, ferrying the dead and injured to Pokhara and Kathmandu. With the avalanche risk lower and the melt underway, hikers were marching back to Besisahar. Mike and I left our bikes and most of our gear at the homestay, and set out on foot, aiming to get as far as we could in spite of the deluge. We began tramping over the hills north-west of Manang, plastic bags tied about our feet, whittled branches for hiking poles.

The trail was empty save for half a dozen men, plodding, disconsolate, hungry-looking. Two had the red, teary eyes of photokeratitis, snow blindness. Most had been stranded by snow drifts that had swamped teahouses en route and they had only managed to backtrack now. As the final hiker passed us, he paused.

'Over there, you see?' He nodded. I followed his eyeline. 'A body.'

The hiker trudged on, leaving Mike and me staring at a figure lying by the river. There was a red rucksack by the head, and legs, fuzzy with snow.

A moment passed before I heard Mike. 'What do you think?'

I was thinking about a medical adage, a pitiless one. 'You're not dead', they say, 'until you're warm and dead'. Hypothermia preserves brain function and even those dragged from icy lakes or avalanches can sometimes be revived long after resuscitation would otherwise be fruitless. Maybe he'd died days ago, when the blizzard struck, maybe not. We had to check.

When we arrived at the man in the snow, I saw that he was a Tibetan lama, his shaven head pillowed on a red rucksack. In part, there was a look of serenity in the list of his head, the lay of his legs, but I could see something violent in him too – his eyes were open, pale, and glazed with ice. His left arm was bent at the elbow, his hand midway to a fist, grasping at air.

'Jesus,' I heard Mike say, over my shoulder. 'He doesn't look real.' And there was that too – the shiny skin, the waxwork quality to frozen flesh.

I knelt down and went to hold his hand.

The first time I watched a doctor confirm death I was a

fledgling medical student. The dead man was old and thin, his face grey and slipping from his bones. A few of the man's relatives were present, gathered tearily but now suddenly attentive, curious about what happens next in the unfamiliar aftermath of a person's end. I understood the routine: the doctor would check pulses, reaction to painful stimuli, response of the pupils to light. He would listen for heart sounds and perhaps say something soothing to the bereaved. But he surprised me. First, he leant forward, picked up the man's hand, and for a few seconds he held it in pointed silence. I glanced at the relatives, glad to see they hadn't read this as a Lazarus-style comeback. Then, the doctor went through the motions. The gesture seemed rather a soupy one to me, at least until I considered his intention. Holding his hand wasn't for the relatives' benefit, although I think they actually appreciated it, nor for mine, and nor was it a mark of respect for the dead man. I think, now, that it was for his own. That defining human gesture lent weight to an act you may perform often as a junior doctor, a ritual that risks becoming hollow and normalised. A gentle prompt perhaps: *don't become inured to death*. On many occasions since, I too have held a patient's hand when confirming death. It helps me remember, however long and turbulent my shift, that this is less a dead body than a dead *life*, the full stop to a singular story. But when I touched the cold flesh of the lama's hand, his tendons and muscles were frozen rigid, his fingers clawed. So I left his hand there, unheld, in mid-air. If the sceptics are right, if human beings are mere biological machines, he seemed to prove it. He looked more switched off than dead.

Our feet crunched on. We picked over streaks of fresh snow slip, ragged like bleached coral, and here I noticed footprints – whoever made them had been marching the other way. Perhaps they'd been made the night before, the snow memorialising a man's final steps.

Beyond High Camp, 5,000 metres high, the snow was too thick and the risk of an avalanche too great to continue. We gave up on completing the circuit and trudged back towards Manang, our wooden sticks poking tunnels, glacial blue, into the snow. Crags were beginning to show above us now, and the sky was blotted with Himalayan vultures. Sunlight slowly roused the valley, and there was a renaissance of colour and contour. Shrubs were emerging from the snow and scenting the air, and the silence, once so creepy, was swapped for an undertone of meltwater.

Fewer choppers flew overhead now and a week after the snowfall began there were only a few people in Manang who could offer details of events higher up. Days before, a cyclone named Hudhud had unravelled in the Bay of Bengal. As sea winds wheeled to over 100 mph, the Indian police had evacuated 400,000 people near the coast. The weather front muscled up the map, but bands of hikers, with no inkling, had set out to clear a pass they would never find. In dense droves of snow, a few had backtracked and survived, squinting after the forms of porters and mules. More than five hundred people had been rescued from near the pass, scores with frostbite severe enough to need amputations, but it was the fate of the 43 people who died that day on Thorong La that made the disaster international news.

We stopped in a teahouse on our ride back and sat near a group of young British backpackers. A teenager was looking into his iPhone. 'Hey, Jack wants to know if I've seen any dead bodies, ha ha ha!' He was joined by his friends, all guffawing helplessly. I tried to catch Mike's eye, but he was staring at the table.

The road down had been lost within a huge avalanche but already porters had dug a trail and we wheeled our bikes through, the ice on each side overwhelming us, as deep as an

Olympic swimming pool and packed as hard as stone. It was impossible to imagine how anyone could survive such an engulfing. And it was disconcerting, somehow, that the Himalayas looked more beautiful than ever: the high rock faces sheeted with unseasonal snow, the blue sky framing yellow larch and criss-crossing vees of rock.

*

Back in Kathmandu, I plotted my escape. The date of my flight from Mumbai to Hong Kong left me with a few weeks to meander south through India. I left Kathmandu, climbing above the smog and out of one valley and into the next, where a long string of cars was locked in place through the obstinacy of two bus drivers. Rounding a corner, neither had made the merest of concessions. Now the grunting faces of their machines were millimetres apart and someone was going to have to reverse, but by the time that point was conceded, several dozen motorists had done what the drivers here did with professional endeavour: closed all gaps. I squeezed past the cars and should have had the entire downhill to myself, but vehicles in the other lane, sensing an opportunity, were using mine. They veered into my path, flashing their lights – the international symbol for 'I'm about to be a moron: cope with me'. Not my country though, not my rules. I swallowed my road rage.

I coasted back down to Terai and crossed the border into India, passing a touring cyclist, a young Indian, on the way.

'Hey, you got any weed?' He looked painfully hopeful.

'Sorry, man.'

He sighed and said, 'I worked through my stash'.

We wished each other good luck and I held out my hand for him to shake. He looked uncomfortable, but reached out anyway. And then I could see why: he had an extra thumb – preaxial polydactyly, if you want to be fancy – a limp, useless appendage hanging out with his functional ones. A few years ago, Akshat Saxena from Uttar Pradesh became the world record holder for the highest number of digits. He was born in 2010 with seven digits on each hand and ten digits on each foot. Extra digits are more common among Indians than any other nationality. This seemed fitting somehow, something to do with India feeling like an exciting and excessive place. Or, as A. A. Gill put it, 'Whatever it is you're looking for, India has it with six arms on.'

I was now riding through the state of Uttar Pradesh, India's most populous. And fittingly, that's how I remember it, as a muddled vision of people: boys in early-morning cricket games in the dusty spaces on the boundaries of towns; a man carrying a whole bed on his head; women overfeeding me *roti choka*; a circle of children bewitched as I slathered on sun cream; and waves of women in black niqabs, always far from the Muslim men with their dyed-red beards. I felt the impact of the crowds too: the sun dulled by the chalky haze of pollution; litter burning by the road; an acrid fog of melting plastic. I passed through nowhere that I remember with any clarity, and slept often in police stations that smelt of incense, guarded by officers with bayonets who slept on the stone floor. I was never turned away, and was always fed and watered.

When there were no guesthouses and no opportunity to rough camp, I occasionally approached police stations, temples, churches, mosques, even hospitals or schools, asking if there might be a quiet corner in which I could sleep. It's the privilege of a western traveller – in South Asia especially, but in much of the rest of the world too – that you are trusted without question. Try rolling up to a school in the UK, scratching your facial hair and asking if you could shove a bivvy bag on the football pitch for the night. 'In Nepal, tourists are God, you understand?' a hostel owner in Kathmandu had said to me.

When I replied that I did, he looked at me in an intense, imploring way. 'And if you see one of us Nepalis in England, will you help him? Will you love him?'

As I headed south, somewhere in Madhya Pradesh, the neighbouring state, a group of kids on bikes drew up. One boy took the lead.

'Please come to my school.'

'Why?'

'For children, for looking at you.'

'But the children are always looking at me.'

'Do you like me?'

'I like you.'

'Then give me your bicycle.'

They giggled and muttered a plan for their next question in Hindi.

'You know you're in India, right?'

I told him yes. He spoke again to his friends in Hindi, maybe: 'Yeah, he knows.'

I liked the idea that I could have got here by accident. Shit, I was wondering why that guy with India written across his shirt wanted to see my passport.

The boys peeled off and, an hour later, three men on a motorbike cut in front of me, forcing me to stop. One leapt from the bike and asked where I was from. When I answered 'England' he yelled 'Boris Becker!' and mimed some tennis serves until I told him that no, Becker was German. He did one last despondent serve anyway.

I arrived at last into Mumbai and, with most of the guest-houses full, I made do with Hotel Delight. Either someone had a sense of sarcasm, or the delight referred to checking out. It was a place of sour-faced patrons watching loud TV, there were shoe prints on the toilet seat, and I left with a tell-tale line of bedbug bites on my arms.

You could quite easily feel delighted in Mumbai though, with the right frame of mind. I was just out of the hotel, in Colaba district, when a man tried to sell me a balloon bigger than the door. Another sidled up to me tootling a flute, fifty or so instruments flapping from his jacket: 'Flute, sir! Flute!' A book vendor took his turn. Girls spieled ropey English, begging at point-blank range, skimming my arms with their fingers. From somewhere behind my shoulder, I heard 'Postcard? No? Hey man, I got some great dope, you wanna get high?' A Bollywood casting agent offered me a bit part in a film. A man stuck something metal in my ear before I could protest and showed me the wax he'd scraped out, but a policeman shouted at the ear-wallah and he scarpered. I let Mumbai happen to me, zoning out, gazing into the middle distance, people popping up in front of me, like I was on the edge of consciousness. Instinctively, you might raise your guard, blot out the cajolers, but then you'd miss the best bits, the invitations to sit quietly with men and drink tea, talk about cricket or the news. Sometimes, you've just got to take an ear picking.

That afternoon, I saw a man, begging, with elephantiasis, a parasitic infection causing lymph fluid to accumulate and the legs to swell massively. My mind wandered to the other outsiders here. The physically disabled could be very visible in India – many hustled for donations in the streets – but I wondered about those with more closeted disabilities. What did India's bustle mean for the mentally ill, for the withdrawn? How did it feel to be lonely or outcast in Maharashtra, a single state with almost double the population of the UK?

With a head full of questions, I arranged to visit a mental health rehab clinic in Mumbai, and when I arrived at the indistinct building on a street wild with traffic, everyone was expecting me. Twenty voices: 'Hello Stephen!' I was told to linger at the back while yoga jump-started the day. When that was done, everyone was given a drum and asked to describe their mood. Today, some were happy, some sad, a few anxious, and one or two couldn't muster an adjective, or couldn't narrow it down perhaps. I could see that a few had learning difficulties. There were only two women.

When the group split up, a posse rushed over to greet me.

'What's your blood group?' Kumar was hungry to know. He looked like he might have fragile X syndrome: Kumar was big, with learning difficulties, poor coordination, large ears and a heavy jaw. He was also bubbly, child-like and prone to capricious moods.

Rohan, who stood beside Kumar, was a balding, bipolar, middle-class man with a gentle manner, whose manic episodes involved wild spending sprees in dance bars.

'Why you asking him these things, huh?' he said to Kumar, appearing a little embarrassed for the group.

'I'm B plus, what about you?' I said.

'I'm O plus!' said Kumar. Rohan shook his head and let himself smile.

'What's up?' said Kumar. 'Do you know Johnny Depp? Do you know Harry Potter and the half-blood prince?'

I spent the afternoon chatting away with members of the group on their own. Ankit was in his forties and, the most self-possessed of the group, often helped direct the day's activities. He came from a well-to-do household of maids and cooks and had been diagnosed with schizophrenia in his late twenties. His father didn't mention this to Ankit's prospective wife when he'd arranged Ankit's marriage. Not long into the relationship, his new wife found his anti-psychotic medication, there was an argument and Ankit decided to cut back on the tablets.

'I had this abnormal behaviour and whatnot. Muttering to myself also. My wife went to my psychiatrist behind my back, he told her everything and suggested we get a divorce. Three

1. Tea flows, the chat … not so much, but an afternoon of broken English, ropey Arabic and inscrutable mime beats being alone for my thirtieth birthday in Syria.

2. A puncture in Egypt draws the crowds. Being trailed by the police had occasional perks, extra hands to lever tyres was one of them.

3. Along with culverts, only the odd acacia offered shade from the testing Saharan sun in Northern Sudan. You take what you can get.

4. Two angelic children in Ethiopia who asked for a photo, and who will both, in a few moments, fire at me gleefully with slingshots as I scream for mercy.

5. The road to Addis. Nyomi accompanied me for much of the journey through Africa. She made friends easily: a quality to be treasured above all others in a travel buddy.

6. Getting a ride across the Omo River from Ethiopia to Kenya in a dug-out. No border post here. Beyond: the Ilemi Triangle, a dry, thorny, wind-wracked badland, claimed by three nations.

7. Nurses weigh infants in a mobile medical clinic run by the NGO Merlin in Turkana, Kenya's northernmost, vast, marginal county, home to a shifting population, scant infrastructure and plenty of swirling sand.

8. In Namibia roads run to vanishing points but at least you can see the elephants in plenty of time

9. Riding across the side of a young basalt volcano in Malalcahuello-Nalcas National Reserve in central Chile, where 'Boiled to death by lava' was added to my mental list of Stuff To Get Anxious About.

10. Crossing the Andes, this time on Paso Pircas Negras, one of ten occasions I crossed the border between Chile and Argentina. It was hard to resist the strange quiet of these high desert passes in particular, and the sense of vulnerability that remoteness can bring.

11. Snacking, near Paso de Sico, northern Chile – an essential ritual of cycle touring. This was my fourth breakfast.

12. Camping with Nicky on the Salar de Uyuni in Bolivia: cold, high, otherworldly; the magic of the moment fading with each guttural snore from the vicinity of Nicky's tent.

13. Riding from Abancay to Cotahuasi was one of the toughest sections I rode in the Peruvian Andes. The trails were surfaced in fist-sized rocks and wandered to over 5000 metres above sea level, and my bike was laden with days of food and water – all of which brought a special pain to the altitude sickness.

14. On the road between Cuenca and Ambato in Ecuador, I'm violently ambushed by cuteness personified.

15. A sequoia in California, a very big tree – in case you assumed I was very small.

16. For kids living in the bloated rubbish dump come slum of *Bantar Gebang*, on the edge of Jakarta, the stench and decay don't quell the instinct to find fun.

17. A Swiss doctor looks out on the floating village of Picharkrei from the Lake Clinic which bobs on the rising and falling waters of the Tonle Sap, Cambodia.

18. Taking on fuel in Yangon while demonstrating that being away from Britain for several years eventually makes you capable of eye contact with strangers.

19. In Myanmar, I walk my injured bike along, approaching the Indian border post at Tamu-Moreh where I hope to find a mechanic better than me.

20. Tea pickers laughing near the village of Matelli in West Bengal, possibly at the state of my clothes and attempt at facial hair.

21. Terai is the long flatland of southern Nepal and northern India, where cyclists abound. In the long sunny evenings I'd race the kids, but winning is tiring, so I let this old fella off.

22. Mike takes the lead, climbing on the Annapurna circuit in peak dry season, with no sense of the catastrophic weather to come.

23. A day after heavy snowfall and a fatal disaster, we set out to find High Camp on the Annapurna Circuit, Mike ahead again, the uncanny silence of the mountains broken only by the odd rumble of a distant avalanche.

24. Hong Kong protests and the financial district in Admiralty becomes 'the village'.

25. The surface of Lake Khövsgöl in northern Mongolia, on the verge of Siberia. The ice is a metre thick in winter, thick enough to support trucks, let alone me – though this is something that bears repeating to yourself, loud and prayer-like, as you cycle across the creaking ice.

26. Early morning repairs on Lake Khövsgöl. The temperature had slumped to −38°C in the early hours and my painfully cold hands were not well suited to the task. There were some bad words spoken and this scene was markedly less tranquil than it appears.

27. Climbing out of Uliastay in Mongolia, this herder reached inside his deel and, with the flourish of a magician, produced a cute goat.

28. A road through the Tian Shan mountains in Xinjiang, western China, near the Kazakh border. Doff your cap to the Chinese engineers who managed this.

29. A mother and daughter outside their yurt in Tajikistan near the shores of Lake Zorkul. Life at over 4,000 metres above sea level, sharing the slopes with yaks and wild sheep.

30. The Pamirs – taking a second to admire a tributary of the Pyanj river and the landscape beyond. A new range rises across the river and fans out into Afghanistan: the Hindu Kush.

31. A street scene in Mazar-e-Sharif, northern Afghanistan, a city of notable diversity, where I can be seen failing, yet again, to blend in.

years married – it was over. I'd tried to be good to her, you know? She was very upset with me. My family get many proposals. I told my father to be honest next time. They have all said no after that. But that's why I like it here in rehab, nobody judges you, there's lots of understanding.'

Rashi was the head of the rehab programme, and we met in her office.

'A lot of families say "you must pray". There are lots of superstitions in India, and many patients have already seen spiritual men who claim to help. People sometimes think their relatives are *deliberately* wrecking their lives. And relatives are embarrassed. We've had patients tied up or locked out most of the time, only coming home to sleep. It's tough because there's only one private hospital with a psychiatric ward in the whole of Mumbai and the public hospitals are disastrous. They will only take patients who have family members of the same sex who will stay with them around the clock. And judges decide if a patient needs treatment, not doctors. Patients are often abandoned, there's no reintegration, no friendships.'

The Times of India had recently published an article for National Disability Day. An accompanying photo showed famous Indians with each of the six categories of disability, except for mental illness – it's in the margins of the margins. To get a disability certificate you have to be '40 per cent disabled', and, as Rashi noted, understatedly, 'It's hard to quantify mental illness in this way.'

Poor mental health was the undercurrent to much of my work in the emergency department. It was there at the beginning of my shift, in the shouts of anger or fear from the mental health assessment room. And it was there at the end of my shift too, as I wandered out of the hospital, past a woman drenched and shivering beside paramedics, having leapt from Westminster Bridge. And it was a spectre throughout, in the sheer numbers of mentally unwell patients I saw, whether or not they had come to A&E because of a decline in their mental health. Your mood and feelings will shape your experience of illness of course, but, more importantly, people with a mental illness diagnosis have a much greater chance of suffering a range of other illnesses too. There are numerous reasons for this. Such patients might be late to seek medical attention, or be non-compliant with treatment, or have lifestyles not conducive to good health. And they are discriminated against within the healthcare system; something I, as their doctor, had to guard against.

For patients brought into A&E seeking help for mental illness, or found acting bizarrely by friends, family or strangers, one of my initial tasks was to assess for problems distinct from 'mental illness' that might have triggered a change in someone's mental state: low blood sugar, a head injury, a nose full of ketamine (not that they'd tell me). Drug withdrawal, especially when crystal meth's involved, can turn into intense paranoia. Provided the patient wasn't intoxicated, or deemed 'medically unwell'* psychiatry would swoop in for a more thorough review, where the patient's various worries, safety and sadnesses could be combed over in more depth, alongside the facts of their life.

And the facts, it turned out, were vital. Within the swirling sphere of mental illness, there were patterns. Immigration, addiction, social isolation, homelessness and poverty are all

* I've always found the language we use around mental illness particularly confusing, not least because it often revolves around the notion of Cartesian duality, i.e. detaching mind problems from body problems. And if you have an issue with rigid categories in general, as perhaps we all should, then you'll find defining any mental illness a particularly messy business.

associated in one way or another – for the latter four at least, it was often hard to know if they played a part in causing mental illness or if they were the result of it. Indeed, both could be true. There's a clear association between mental illness and childhood trauma as well, though this was not something I would generally discover in the short time I spent with my patients. Sometimes, it wasn't entirely clear if this was 'mental illness' or not. A young drug user in a hostel. People are watching him through the floorboards, he tells me, stealing his stuff and whispering threats to kill him. He's being followed. Is he delusional and hallucinating, the hallmarks of psychosis? Problem is, for homeless addicts, such a situation could be their reality. It's not paranoia, as the saying goes, if they really are out to get you. I'm sure hostels can be intimidating places.

Perhaps I'd been expecting that mental illness would be very different in India. It wasn't, not really. I'd only been granted a snapshot of course, but even so, the similarities were blatant. Here too, mental illness had been pushed down the list of priorities, and for the same reasons, I suspect – because it requires time and resources, because it's considered uncomfortable and tricky, if not hopeless, and because often, those affected don't have a loud political voice. There were some particular intersections in India, with religious beliefs and arranged marriage for instance, but otherwise the same vicious circles were at work: mental suffering can make it hard to hold down a job, can feed feelings of insecurity and guilt, can make it tough to relate to others, and can make people vulnerable to discrimination. Cue unemployment, shaky living arrangements, dependency, low self-esteem, poor relations – all feeding the despair and hopelessness. Spin, spin, spin.

I find it heartening to see mental health services staffed by dedicated advocates for the mentally ill, and it's tempting at such times to conclude that we've come a long way from the cruel age of asylums. But then what's orthodox in one era, enlightened even, may be seen as unjust or cruel in another, such are the tides of moral relativity. It's inevitable that there are ways we treat the mentally ill today that will be seen as abhorrent a hundred years from now. Perhaps it will be the stigmatisation of a mental illness diagnosis, or perhaps pharmacological treatment will be branded akin to punishment. Who can say?

India had asylums too, many of them set up in the late eighteenth century, during colonial rule, including one here in Mumbai. The Bombay Lunatic Asylum was divided by the British into areas for high-class Europeans, 'females of all caste and colour', European males, and native males, the last of these housed in a wing in a miserable state. Class discrimination and racial segregation were implemented together, inside an institution that was already divisive in essence, segregating the mentally well from the mentally unwell.

I thought back to Parbati, deformed with leprosy in Anandaban Hospital. Historically, mental illness and leprosy have been similarly stigmatised. The French philosopher Michel Foucault wrote at length about such connections. Both 'lepers' and 'lunatics' were more often segregated during times of crisis and fear, where populations worried more than usual about vagrancy and outsiders. Strangers are quickly rounded on when we're struggling with our own sense of belonging. Foucault also suggested that the segregation of lepers in leprosariums was the precursor to other dividing practices, of which the asylum was a later one. In fact, the same buildings were sometimes used, and many leprosariums in Britain later became asylums. Today, many of the old asylums in Britain have been demolished, while others have been converted to modern mental health services, and others to blocks of luxury flats with walk-in showers, mock-Regency panel doors and on-site solariums. In 2003, the local council stepped in to buy St David's Hospital in Carmarthen, one of many old asylums across Britain, for three million pounds, after speculation around government plans for it. If the rumours had become reality though, it would have been the continuation of a legacy, not to mention a semantic coincidence. An asylum for asylum seekers. A walled-off domain for the next maligned pariah: the refugee and the migrant.

*

On 15 October 1923, in Bombay, as Mumbai was then, six garlanded Indian weightlifters posed next to bicycles with drop-down bars and Dunlop tyres, ready to launch themselves into an adventurous migration. All were young, moustached and carried satchels on their backs. For their send-off, friends at the Bombay Weight Lifting Club had gathered musicians who were loud, if not particularly musical, and with added cannon fire, their farewell party could be heard for miles.

The men were too bashful to disclose their real plan, which was to cycle from India, around the world and home again. For a start, cycling around the world wasn't something Indians were noted for doing and when they'd tried to cadge money for the expedition from wealthy potential benefactors, they were laughed out of the room or taken aside and counselled on the treacheries of the road; wild beasts and mythical creatures were summoned into possibility, giants were said to stalk the deserts. It was easier to tell their friends and families that they planned on pedalling as far as Persia.

Their other goals were simple, candid and far more laudable than the wild egotism of Fraser a generation before. In the account of their journey, one of the bunch (it's not entirely clear which) writes:

> Young as we were, we were fired by an intense desire to carry the name of our country – Mother India – to the far-flung corners of the earth ... we wanted to know the world more intimately and to acquaint the world with India and Indians ... If there were more intermingling amongst nations of the world, if there were more intimate interchange of views and cultivation of that spirit of international brotherhood, we would have less of wars and God's world would be a place certainly worth living in. Is that not a sufficient excuse for our enterprise?

Appropriately for their hobby of weightlifting, they were seriously loaded. They lugged numerous maps and charts, a doctor's dispensary on a miniature scale, and a Primus stove that ran on oil. Within the team, the men had distinct roles: one was navigator, another, cameraman. One of the six, Pochkhanawalla, was granted the role of expedition medic on the basis that his brother-in-law was a medical practitioner and had 'initiated him into the mysteries of the profession within a fortnight'. Uh-oh.

Not long after the expedition party departed, the labels came off Pochkhanawalla's medicines and pill boxes fell open within the capacious medical chest. Tablets were now indistinguishable, as were the hypodermic syringes of vaccines and bottles of iodine, tannic acid and glycerine mixture. Pochkhanawalla abandoned the unidentified portion of the medical chest on the roadside. But, 'scarcely had we mounted our bikes when a loud thud was heard – Pochkhanawalla was on the ground, nursing his ankle ... We administered to him first aid under his own instructions and proceeded.'

With their accident-prone medic back on his bicycle, they moved west into the Persian desert where their cycle-tube joints melted in the heat. Wind blew sand into their Vaselined faces, where it stuck. At night, they preferred the cold, open desert to the occasional caravanserais, which were often hotspots for lice and insects.*

The men's money ran low early on, and, after debating schemes about how to renew it, they decided that their weight-lifting experience might help. In Persia, they were performing to audiences: bar bending, fencing, boxing, jiu-jitsu, breaking stones on their chests and dragging a car with five passengers by a rope held by one man's teeth. To avoid road tolls, which furnished corrupt officials, they had firmans, letters accumulated from important and distinguished people they'd met en route, which instructed people to help and look after them. They had scribblings from the governor of Duzdab, the Persian minister for war, and, later in their travels, an autograph from Mussolini himself, which they'd secured during their stay in Rome: 'It's effect was electrical.'

On 1 July 1924 the men began the journey from Baghdad to Aleppo, 549 miles over sand that would take 23 days. They put pistachio nuts and pebbles in their mouths to stave off thirst, and a couple of onions in their hats – hearsay suggested it would keep them cool. In the intense heat, punctures were tough to fix as the glue didn't set, so they poured precious water onto their tubes to cool the rubber down. When they lost sight of the Euphrates River, they ate the onions in their hats. Lost, they grew unbearably thirsty and at last they stalled.

> One by one resigned himself to the worst fate. We sat in a half-dazed condition for full 36 hours. We were in a semi-conscious condition. We did not make any effort to move. We simply waited to see one companion after another pass into Eternity.

---

* Every good bicycle adventure needs an infestation from time to time. I picked up fleas in Ethiopia, you'll be happy to hear.

But incredibly, in this trench of hopelessness, soldiers appeared – it was the Foreign Legion. Water was provided; the men could bike on.

They crossed Europe in the interwar years when much of the continent was dilapidated, creaking under massive inflation, full of half-completed buildings, poverty and prostitution. ('There is a proverb "see Naples and die". We have seen Naples and we do not understand what the proverb means.') But the men were committed anglophiles and they arrived into London waxing lyrical about a cosmopolitan city, of rich variety and Rolls-Royce cars: 'Incomparable London, wonderful London, London which seemed to captivate our hearts and throw out our bodies!'*

At the Cycle Tourers Club in London, the Indians were hailed as heroes, with a few creative journalists writing that, back in India, the men came across lions and tigers in the street. But sadly, Britain didn't always love them back. 'Often we found some of the Englishmen too haughty to condescend to talk to us. Dark skin, tanned by exposure to the sun, was the only reason so far as we could trace ...'

---

* Such wonder has been pathologised, of course. Stendhal syndrome involves dizziness, fainting, confusion and even hallucinations when an individual is exposed to an experience of great personal significance, particularly viewing art. The 'illness', also wonderfully named hyperkulturaemia (as in 'too much culture in the blood'), was named after the nineteenth-century French author Stendhal, who described his experience with the phenomenon during his 1817 visit to the Basilica of Santa Croce in Florence. Famously, he wrote: 'I was in a sort of ecstasy, from the idea of being in Florence, close to the great men whose tombs I had seen. Absorbed in the contemplation of sublime beauty ... I reached the point where one encounters celestial sensations ... Everything spoke so vividly to my soul. Ah, if I could only forget. I had palpitations of the heart, what in Berlin they call "nerves". Life was drained from me. I walked with the fear of falling.'

During an interview, one newspaper reporter asked 'Do you mean to say you Blacks could do it all?'

Once they'd sailed to America, their race was again a hindrance – at the time, degrading medical inspections were the norm for non-Europeans. To many Americans, India was a wild, mysterious place, and had a reputation that could be exploited. Hustlers with a dark complexion would put on long robes and turbans and profess to be Indian fortune tellers, charging money for their magic, or passing themselves off as genuine 'Indian dancers'. When confronted by these con artists (who couldn't, of course, speak Hindi) the men complained, pleading with the scammers to revise their nationality.

When they entered Saigon, they were detained and fingerprinted: the French colonialists had strict rules for Indians and Chinese, while other nationalities could pass quite freely. Only a decade before, Indian soldiers had fought in Flanders to protect France. When the men wrote to a local newspaper complaining of their treatment, they were swiftly arrested and almost deported to Singapore.

Finally, after more than four years, the men arrived back in Bombay to be 'garlanded by the people until we were buried in flowers'. How did they feel on concluding their adventure?

'We felt proud we did our wee bit for Mother India, whose illustrious name we carried into the nooks and corners of the world, where we showed sons of Mother India were as able, as enterprising, and as courageous as the children of any other nation in the world.'

As able, not better. At the time, colonisation was often framed as an act of charity and it moulded the borders of the world as much as the minds of its citizens. And while I love this story of the intrepid Indian cyclists, I can't help feeling gutted that they had to strive for par.

Part Four

HONG KONG TO CALAIS

Leave the door open for the unknown, the door into the dark. That's where the most important things come from, where you yourself came from, and where you will go.

Rebecca Solnit, *A Field Guide to Getting Lost*

11

Plexus

Given everything that came later, it's easy to forget that Hong Kong didn't pique my interest once. It was a cheap flight away from India, a staging post, not somewhere I was particularly inspired to see. I expected a city of consumers, busily consuming. A city tediously addicted to profit and commerce. A glittering spectacle sure, but no soul, no fire.

I was wrong, of course, but especially in late 2014, when Hong Kong was anything but dull. Ten weeks before I arrived, the first of many protests had sparked out of rumbling anxieties over constraints on democracy and the growing dominion of the state. The protests escalated fast, and far beyond expectations. The movement called itself Occupy Central, and it did so: thousands of protesters were camping en masse in cheap tents. More than 100,000 had taken to the streets to demand the chance to elect Hong Kong's next head of government in open elections without Chinese oversight. The then Chief Executive, CY Leung, had recently stated that it was unacceptable for his job to be chosen in this manner because 'doing so would risk giving poorer residents a dominant voice in politics'. Kerpow, plebs.

I took an afternoon to wander about Admiralty, part of the central business district. 'The village', its new nickname, felt apt; there was a clear mood of togetherness here. Thousands of tents were jumbled together on roads and pavements, with partitioned areas for younger protesters to take study breaks. Gas masks hung from deckchairs where seditious old folks sat, supping ginseng tea. An organised team of supporters handed out noodles, collected litter and recyclables, and sprayed each other with water to quell the heat of the day. A man was quietly reading *1984*, and behind him photos from the protests sheathed concrete bridge supports. One stood out, an aerial shot of yellow umbrellas overlapping like scales on a snake.

But these were the twilight days. The fight was dying and the most cheering banners had been lost in gusts of police enforcement: 'You may say I'm a dreamer but I'm not the only one' was getting torn down. 'We'll be back' affirmed red paint. Still, impossible not to be heartened by the dualism: billion-dollar skyscrapers swarming with five-dollar tents; colour and playfulness beside right angles and steel; hopeful rebels defying imperious demands for order and obedience. TV crews weaved through the tent-city because the David-and-Goliath-ness had enthralled the world's press too.

I was staying away from the action, on Lantau, the largest island in the region of Hong Kong. In the small town of Mui Wo I hung out at Tom's café. I liked Tom's, it was an odd mix of Hongkongers, holidayers and well to-do economic migrants,* with a sprinkling of eccentrics. There was a mum with a kid called Acacia (there was no sign of her brother, Wigbert. Or, for that matter her sister, Ptarmigan). I was joined at my table by a Scottish doctor who mixed general practice with Chinese medicine and charged big money for a consultation at her clinic in central Hong Kong, and then by a man with a plumy British accent who advised large corporations on how to structure their business in ways inspired by the Kenyan Maasai. He asked about my journey, so I gave him the bones, and he said

* 'Expatriates'.

'Shit a brick! Shit a bloody brick!' I have no idea where these people come from, but Hong Kong seems to encourage them.

Time to go, so I hopped onto the ferry to Hong Kong Island. The sun was low and amber; reflected in the windows of superstructures, the effect was a city-wide inferno. A helicopter dallied and then rolled from view. On the pier I could see a collection of men in Santa outfits, stumbling along, arm over shoulder, latecomers for the international event of Santa-con where participants dress up for a pub crawl. I thought about how, not so long ago, this was all a lonely scatter of rocks, a fishing and farming village, before rapid expansion in the 1950s, a financial flourishing, and now a man throwing up half-digested dim sum into his Santa hat. And I guess that's progress.

Right then. China. I smoothed out my map on a bench in the ferry terminal, and stared into a noodle soup of roads. Villages abounded with the rare beasts of the Latin alphabet, x's, q's and z's, their English translations terrifyingly alike. When China took up one sheet of A4, my beeline to Mongolia looked uncomplicated enough, but zooming in, my troubles were magnified too. My first task was to negotiate Guangzhou, easier for westerners to utter than other Chinese cities perhaps (I'm looking at you, Shijiazhuang), but harder than most to round. It's a hubbub of 14 million people – as many as live in London, Birmingham and Manchester combined – itself within a conurbation of 44 million people, the Pearl River Delta, which has been the most economically dynamic region in China for as long as I've been alive.

Shadows fell over my map. I looked up. A group of young, besuited Caucasian boys were huddled around me.

'Hey, how are you loving Hong Kong?'

The accent was American, but with a twang I couldn't quite locate. Perhaps they'd got lost on a school trip.

'Great, thanks. What are you ...' I trailed off, noticing

their silver name badges: Elder Bingham, Elder Jacobsen. Shit. When did Hong Kong get Mormons?

They were happy to explain it. It turned out that they were here to learn Chinese for six months before they hit the fertile (70 per cent atheist) ground of mainland China, working as 'English teachers' and doing some clandestine (and illegal) proselytising on the side. In the 1920s, Chinese slogans proposed 'one more Christian, one less Chinese', but in recent years churches had shot up over much of the nation, and by 2030 China might have more churchgoers than America. (There is no shortage of mad statistics on China, simply because it's so populated. The country probably has more polo players than Britain and eats more baguettes than France.)

Fearing I was about to get lectured, or overpowered and baptised, I made my excuses and jumped onto a ferry to the Chinese mainland. Shuffling off again at Zhuhai, I blinked up at a signpost, then into my map. I'd allegorised the first symbol in the names of the coming towns so that I could pick them out, but I couldn't see a man-with-box-for-a-head attacking a giant spider. I studied my map again and rechecked the signpost. Huh. No alien-with-a-scimitar either.

But then nothing was as it seemed. What I'd taken for minor roads were clamorous four-lane affairs and these were dwarfed by eight-lane expressways, green on my map, themselves outdone by the reds and whites of even meatier thoroughfares, aortas of streaming metal where a cyclist would be flipped around like a red blood cell. I abandoned State Highway 105 for a road that arched by a row of epic warehouses, begetting that catchphrase 'Made in China'. I slept in a tent sewn together here, and my bicycle bell was dinged here years before I picked it off a shelf in a small English market town.

Night fell over the metropolis and I plunged deeper into an endless phosphorescence. Guangzhou was unfathomably big, but it was just one of over a hundred Chinese cities that was home to more than one million people. It was one of fifteen megacities too, all marching outwards and upwards in the fastest, largest (and most discussed) urbanisation in human history. More than half of all Chinese are now city dwellers, but there is scope for greater gathering and more development, for ring roads about conurbations to multiply like tree rings. I crossed a bridge that allowed container ships beneath. High-speed trains punched through the dark, and even at this hour cranes beavered away, proliferating the identikit tower blocks. I had an unsettling sense of something runaway here, a dystopian future, where all sense of wildness would be consumed by an endless synthetic sprawl, with no stillness, only toil and speed. But as I cycled from puddles of darkness to streetlight, there was a heady sense of infiltration too. Hip-hop in my headphones helped, with Jehst's laid-back drawl: 'The king blue twister, smash your transistor, it's the High Plains Drifter, that had to resist the, Sickness of the city life, I'm sat by the river ...'

I camped that night in the edgelands, beside a construction site. Eight flyovers latticed my view.

The next morning, men stopped eating as I slunk inside a restaurant in the small town of Xiaoxiangzhen, or possibly nearby Xiangangzhen. I scanned the menu – a scrabble of Chinese script – and wondered briefly whether it was something else entirely, a calendar maybe, and what the waitress would make of me if I pointed to April. In Hong Kong, I'd found an outdated, 1,000-page guidebook for China, but now I found that the language section was unreasonably slim and was divided into Cantonese, Mongolian, Tibetan and Mandarin. Also, it contained no word in Cantonese or Mandarin for rice and no word for noodles. Though it was good to know that if I could suppress the urge to eat for the next three months, I could ask 'Where can I buy a padlock?'

I strained to order breakfast through mime, thick fag smoke like the dry ice of a stage performance. My beef impressions failed, despite – and I know this might sound boastful – *very* convincing finger-horns. I needed the toilet now too, but the book had let me down again. There is no way to mime 'toilet' without offending people, even if you don't make a constipated face and include the farting noises. I stared glumly at the book. Where can I buy a padlock? As a euphemism, it didn't really work.

I settled, in the end, for a bun on the counter, which at least had a label translated into English: 'Best enjoyment in spite of your care. Tasting it still remains so exquisite, the fantastic feeling hovering above your head gives you colourful dream at that moment.' It was okay, for a clump of damp bread.

'Foreigners are easy to fool' goes the Chinese proverb, accurate in my case. I'd mastered some simple greetings in Chinese now, but I stumbled helplessly over others. My age, 34, was a sibilant nightmare, which I pronounced like a drunk ordering samosas. Pinyin, one of several forms Chinese takes in the Latin script, had signs hanging about the letters that indicated phonemic tone and determined the meaning (I'd ditched the idea of learning Cantonese – it had even more tones). My favourite tonal catastrophes were: *shùxué*, which means 'mathematics', while *shūxuě* means 'blood transfusion'.

'Doctor. The patient's bleeding out. What's the plan?'

'Mathematics!'

Guòjiǎng means 'you flatter me' while *guoǎjiàng* means 'fruit paste' – as in:

'I love what you've done to your eyebrows.'

'Oh! Fruit paste!'

Alas, I'm not a natural linguist, though a good short-term memory helps a little, something that was boosted during medical school, where every student is presented with a new linguistic repertoire to master. Medical dictionaries can be daunting tomes, and as a junior doctor I often felt I was drowning in this new vernacular, from long, Latinate names of diseases and anatomical landmarks to countless eponymous syndromes and myriad acronyms. Some terms don't exactly roll off the tongue; sphenopalatine ganglioneuralgia, for instance, is the headache you get from eating ice cream.

There was medical slang too, a way of blunting the pressures of the job (at least, this is the kindest way to excuse it), and this was my first inkling that doctors were as capable of sourness as anyone: not machines, not angels, not always right or good. There have been attempts to compile this playful, sometimes cruel, idiom. In the US, a 'Double Whopper with Cheese' has been applied to obese ladies with genital thrush. In Britain, 'Pumpkin positive' references patients so stupid that shining a light in their pupil might cause their whole head to glow. There's the 'cold tea sign', which refers to a patient with several cups of cold tea by their bed – and implies that they're dead. But then perhaps Hippocrates himself would have approved: 'The physician must have at his command a certain ready wit, as dourness is repulsive both to the healthy and the sick.' Sometimes, patients were in on the joke too. I'd once asked after a patient's medical history, a man with coeliac disease. 'Well, doc,' he said, with a smile. 'I'm glutarded.'

Some medical conditions sound horrible, and they are. Take 'generalised paralysis of the insane', a 'flail chest' or 'fatal familial insomnia', in which tiny, accumulated proteins in your brain prevent you from sleeping properly for the final, harrowing three years of your life. A patient, having sustained a specific pattern of facial fractures, can be said to have a 'floating face'. Such patterns were identified in the nineteenth century by the French physician Le Fort, who dropped cannonballs onto the faces of cadavers and classified the mess made afterwards. Phwoar.

Other terms are less apt or intuitive, and the body can go hideously awry in ways that sound deviously benign. Take compartment syndrome: fairly innocuous, right? Well, not after you see it in the taut and dying flesh it's not. A megacolon is not as great as it sounds, and trust me, you never want to 'cone' if you can help it.*

But behind all the jibes, the horrors, the wonders and the clues, medicine has a poetry too. When I lean over a patient and listen to their chest with my stethoscope, I sometimes ask them to say 'ninety-nine'. What I hear back is coined 'whispering pectoriloquy' and it sounds like a spell. If I hear the rush of blood through a pregnant uterus, I'm listening to 'a souffle', and, without my stethoscope, there's the gurgling of the bowel, onomatopoeically named 'borborygmi'. Medicine, you're told, is a science. And then you're dazzled by its art.

*

Crossing backcountry China can be a lonely business when you're murdering Mandarin, and when your family are pushed beyond China's great online firewall. But then, one afternoon, there was a scrawny cyclist with panniers, resting by the highway. He looked Chinese and he was the first travelling biker I'd seen in the country so far. Some mutual bellowing decided it: we had no common language, so I spread out my map on the ground and we leaned over it, wide-eyed, pointing, like generals planning an advance.

He was heading for Tibet, it seemed. Occasionally foreigners sneaked in, but only Chinese bikers were allowed to ride the

* When part of the brain slumps through the hole in the base of the skull, compressing the brainstem, which turns out to be quite an important bit of you. Coning tends to render you dead.

plateau without being saddled with a tour group, relieved of life savings, and ordered what to enjoy and eat and photograph and historically re-interpret. But for now, we were heading in a similar direction. He reached for his phone, typed and handed it back to me. The screen read 'we are kinsmen'. Then he dialled, spoke into it and passed it to me. I heard a voice from Beijing: 'Hallo! Hallo! He want to cycle with you, okay? You go now, you go together. You help each other.'

We did as we were told, riding off together, towards a bevy of hills crowned with evergreen forest. The cyclist stopped up ahead, took a breath, yelled out over the view. I took my cue and unleashed: 'AHHHHH!' For the next ten minutes we took it in turns, rejoicing in the space and laughing at the echoes. I waved him goodbye that afternoon as he arced away, to the west. If we'd ended the day together, I might have asked his name. And he might have nodded and said 'yes'.

In rural Guangxi, the villages smelt of aniseed. Old men, as attentive as surgeons, stood hunched over marathon games of mahjong. I passed through farming villages, where animals were slaughtered roadside. There were cows with slit necks, making their final moos; dead pigs being shaved and inflated with bicycle pumps; and ferret-badgers, skinned and waiting for customers in a slimy heap, spilling claggy blood over the road edge. Once, I saw a man crouched down, in overalls, holding the blue flame of a blowtorch to the paws of a dog in rigor mortis. He looked at me with an expression entirely befitting a man blowtorching a dog, and so, despite having quite a few questions, I had no desire to stay and ask them. In a petrol station, I bought a can of drink pulled from a fridge stuffed with carcasses. What came out of it tasted vaguely meaty, though because of my debilitating ignorance of Chinese characters the drink itself could have been donkey-flavoured. Roadside markets flashed by, with hunks of animal, cross-sections, dried and sliced and hung up in deep racks. Choose your T-shirt, then choose your goat, lamb or dog.

Meat has a potent symbolism in China, its popularity hitched to notions of health and wealth and male virility, all the more so in light of the nation's recent past. The famine to wrack China under Mao's leadership from 1958 to 1962 was the world's most devastating for at least the last 150 years. Even a state-sponsored optimist would admit to 15 million deaths, and some estimate that it's three times that number. It was, in large part, the upshot of Mao's calamitous Great Leap Forward, and came amid great poverty and dire health facilities, where doctors were often sick themselves and served their patients not in white coats, but rags. To critique the powers that be was a fatal mistake, even as those around you died. Disease bloomed in the starving villages, and there were outbreaks of polio, malaria, measles and hepatitis. As conditions degenerated, so did the strategies to find food. People ate plankton, sawdust and wood pulp to stay alive. They inadvertently poisoned themselves with cassava leaves, anthrax and dead rats. A farmer who negotiated a price for a package of meat at Zhangye railway station in 1960 found a human nose and several ears inside. It is almost impossible to think of a more chilling consequence of starvation than cannibalism, but a mind-shattering tableau in Guanyin comes close. Here, villagers turned to eating the very earth to stay alive, as Frank Dikötter explains in *Mao's Great Famine*:

> It was a vision of hell, as serried ranks of ghostly villagers queued up in front of deep pits, their shrivelled bodies pouring with sweat under the glare of the sun, waiting for their turn to scramble down the hole and carve out a few handfuls of porcelain white mud. Children, their ribs starting to show through their skin, fainted from exhaustion,

> their grimy bodies looking like mud sculptures shadowing the earth.

Some 10,000 people ate a quarter of a million tons of soil in Guanyin, and many were left unable to defecate and had to prise out their faeces with twigs.

*

North of Mengshan, I cycled beneath brooding podiums of limestone, fluted with foliage – karst peaks, once the walls of ancient caves, long collapsed. The karst had brought a throng of sightseers to Yangshou where it was lit with spotlights by night. It was New Year's Eve so I sought out an international crew in one of the hostels. I'd gone almost two weeks without a conversation of any kind, though ironically, I now had the kind of faraway, urgent look in my eyes that made socialising much less likely.

Beside my hostel, there was a bar, and an Australian called Darryl was behind it, because this is the way of the world. Darryl was selling shots from a large jar of rice wine stuffed with a number of grey, putrefying snakes. He insisted on shouting me a shot, which of course led to more. The fun that followed was tainted by a feeling that I had no real friends here, only flushed twenty-somethings, groping each other after beer pong. I missed proper connections, though this was a simple fact of my lifestyle and I knew that it was childish to protest or whinge about it. Go home, or swallow it (and the fetid snake sweat too).

The next day, with a reptilian hangover, I followed the gentle turns of the Lijiang River. Tourists motored along beside me on bamboo rafts with sun canopies, and cormorant fishers strolled the banks, the birds perched imperially on shouldered sticks. As I crossed into the state of Hunan, a sign: 'Welcome to Joyful Dong Land'. Visions of a commune of well-endowed naturists was put to bed by a quick flick through my guidebook. The Dong people were an ethnic group living amid bamboo and misty rice terraces and soon I saw huddles of men, dressed down in navy blue or black, and women, singing heartfelt songs around coal fires.

The police arrived a few hours later. I was sitting in an eating house, drinking tasteless tea and mourning the Indian version. The three officers marching towards me looked severe and obviously about to do me no good. I was quickly led away to an imposing white-tiled police station, by far the largest building in town, where one officer explained the problem.

'Restricted zone. No foreigners.'

Actually, this made some sense. I'd read that China's intercontinental ballistic missile system was located near here, a fact I'd noted with amusement in large, underlined letters in my soon-to-be-inspected and evidence-bagged diary. The 'restricted zone' had not been signposted (at least, not in any way I could understand), there was no barbed-wire perimeter, and it was, quite possibly, the size of Greater London.

'Camera!'

I gave it up. A tall officer began flicking through the images. He had an X-shaped scar beneath his left eye and it struck me as comically clichéd that he should also be the 'bad cop' of the bunch. And then something inside me fell, fast. The Hong Kong protest site, the tents, the anti-Beijing slogans. Fuck.

'You wait here. Work mates coming. Two hours.'

'Who?'

'Special officers.'

Work mates? Special officers? Intelligence? An international incident? Chinese water torture? A hard-labour camp? A public lynching? Something worse? Piers Morgan touching my nipples?

'Hello, sir.'

The special officers spoke fluent English, which I took to mean that I was in all sorts of trouble. The young woman was placid, smiling and petrifying, her colleague was a sour-faced older man, also petrifying. My documents were demanded and photographed in triplicate. I was told to describe my route over a map as the special officers took notes.

'Now we take a look at your things,' she said, with another piercing smile. 'Just looking, okay?'

The search, slow and methodical, was filmed on video camera. They retrieved a mouldy muffin. They learnt that my panniers smelt fungal (I'd been riding for seven rainy days now without a break; I hadn't been expecting guests). Surely, they didn't take me for a foreign spy now? And if not, they were probably just bored, and just snooping. I was further persuaded that this was the case when, inside, I found seven village cops sitting around my computer, opening files, perusing documents and images and looking extremely happy that they had some proper crime to get busy with and no longer had to investigate the theft of Mr Shan's chicken. But they were closing in on the Hong Kong protest snaps that I'd copied onto my hard drive, and getting closer to revoking my visa, or at least denying me the extension I needed. I had to make a call, so I drew myself up and snapped 'Enough! You've seen everything, okay?'

For a moment, I wondered if I'd overcooked it and the silence that followed was edgy. But the female officer closed my netbook and handed it back to me.

'Okay. Take us to where you sleep last night.'

'Take you?'

'Yes. Come.'

We piled into three police cars and I directed the squadron to the river bank. The stony ground on which I'd laid my tent the previous night was photographed at various angles, but the mood had shifted now and the officers were all chuckling at such aberrant behaviour.

Encouraged by this, I drew myself up again. 'Can I go now?'

'Yes, yes, but first you will join us for lunch,' ordered the older special officer.

We drove to a restaurant in Hongjiang, on the edge of the restricted zone. I was served first. Dumplings were mounded onto my plate, fish and pork, tofu and fresh vegetables. A beer met my hand, someone clapped my back. As I reattached panniers to my bike, the police lined up, as if for a school photo, smiling a collective goodbye.

Apparently, my name had not been added to a government blacklist – this was confirmed in Changsha where my visa extension was wonderfully easy to arrange. Mao Zedong had once lived and studied here before joining the Communist Party and there was time to look around the various Mao-based attractions. His massive head was a popular one. Not the old, balding Mao, mid-salute: here was the youthful, windblown poet, a glint in his granite eye.

Back in my hostel, a girl sat down opposite me in the reception. She was pretty, with huge glasses that made her eyes shine. She smiled and typed into her phone before handing it to me. It said 'Age?'

I typed a reply.

'Married?' said the screen. I shook my head.

The next message read: 'I'll never get married', and then 'Will you eat ribs with me?'

I typed yes. She smiled, left, and I never saw her again.

*

I was back on the road a few days later, scouting for a camping spot behind a feeble disguise of bushes. But in Hunan province, what made for a subtle wild camping spot could be a popular toilet too and the ground here was strewn with toilet paper. On the road, an outline of a cyclist. They stopped in the lay-by. A disembodied voice: 'You sleep … here? In these … bushes?'

The voice was measured, but concerned. The kind of tone someone might use when asking, 'And have you harmed the hostages?'

'Maybe.'

'Why you single no double?'

I shrugged, suspecting the answer was that I'd become the sort of person who sleeps in toilets.

'I think,' the rider began, teacherly now, 'you come with me.'

Given the alternative of flatulent late-night guests, I pedalled off behind the stranger. Car headlights brought his figure to life, a young bespectacled Chinese man on a mountain bike with two panniers and a rucksack on his back.

His name was Liyan, and he could find stuff, read road signs, order from menus, discuss the whereabouts of cheap hostels with strangers, order chicken soup without chicken feet, barter, phone Chinese people, joke and generally conform to an acceptable standard of human competence. My only faculty, aside from a goofy grin, was a loud and stumbling confession that 'I can't speak Chinese' ('*Wo bu hui shuo Putonghua!*'), which I brayed at anyone who innocently asked for my name or nationality.

Later that night, we sat eating bamboo shoots and tofu beside a food stall. The owner, a furrowed, twitchy woman, chattered away to Liyan and tossed me exuberant glances as if my presence was the best thing to happen all year. I'd experienced this kind of rapt welcome before, but now I had Liyan to translate.

'She says you have a big nose, but she likes you.'

'Tell her, I like her too. But her nose is much too small.' He translated.

'What did she say?' I asked.

'She said yes.'

'Yes?'

'Yes.'

And we all sat silently for a minute, as perhaps she wondered, like me, how to expand on the international discrepancy of nose sizes.

Liyan had cycled from the south too and now he was heading to Xian to spend the approaching new year with his family. This journey home during Chunyun, the Spring Festival, is supposedly the largest annual migration of people on earth. My own route steered east of Xian, but even so, our partnership looked set to last for a few more days at least.

You enter an uneasy deal when cycling with someone else. Questions simmer up about the compatibility issue. Will they be faster or slower than you? Will they fart into the slipstream or croon Bavarian pop songs? Are they breathing that way to piss you off? Luckily, Liyan was a great companion, and I was glad when, after two days, he suggested we ride together for another five. I say glad, but it was mostly relief – Liyan had become my guide, in geography, language and food, and, thankfully, it was role he seemed to relish. I was a gormless man-baby who was now his responsibility to feed, lead around and put to bed. If he tried to dump me, I'd just tag along anyway, pretending I didn't know what he meant.

Liyan's left wrist was wrapped up in bandages, and in his captivating blend of unsure English and mime, he described an unlucky exchange with a car. It would have been a reminder of a cyclist's vulnerability and the peril of Asian roads had I not seen Liyan cycling. He plunged through red lights, and swanned about on frenzied junctions looking absently upwards, as if trying to recall where he'd left his keys. His injury was beginning to look less like bad luck than a fortunate escape from something much worse. After a few days, I began to take the lead, wondering if I could help Liyan survive China too.

Sometimes, the distance between myself and Liyan felt extreme, a cultural chasm, language the least of it, but if anything united us, it was food. Liyan ate faster than anyone I've ever met, and I took some beating myself. I gave him an exaggerated look when he slurped his last noodle, and for two people with only bicycles and a few words in common, this was the best joke going. But then perhaps, out here, Liyan wasn't unusual in this regard. China felt like a racehorse of a nation, where everyone ate fast, drank fast, spoke fast, got rich fast in fast cities. The government decided, and lo, it occurred. For a perpetually impatient soul like me, one who believes that anyone who walks too slowly in the street should be promptly removed from society, ideally by human cannon, China was a kind of bliss.

We found Yueyang by its neon, a flock of hotel signs in the dark. At daybreak, we found the Yangtze River too, and took a boat across its eiderdown of mist. The river was so broad that it took ten minutes to reach the far side, yet beyond that a shelf of land marked an older watermark, from a time before the Three Gorges Dam, the world's largest hydroelectric project. China only had to relocate 1.2 million residents to accommodate it, which I imagined it accomplished over a long weekend.

The cold was cutting now, the wind bawling from the north, a warning from Mongolia. We cycled past cabbage fields, the sky busy with magpies switching trees. A restaurant was an escape from the buffeting. There was an olive-green bin beside it, the size of a small car, where rubbish smouldered. I stopped, squinted. A man was standing inside the bin. Rattily dressed with a thatch of hair, his face was entirely black from filth and soot, his eyes pale cracks. Smoke billowed from the tramp's clothes as he endured a singeing, and in return, escaped the worst of the cold.

The next day, I found Liyan looking gravely into his phone. 'Many many snow. Twenty centimetres come tomorrow.'

The blizzard arrived on cue, and in thick spasms of snow, Liyan was shouting over the wind.

'Too cold for praetorium!'

I considered this for a second. 'Praetorium' was probably 'tent'. The translation app on his phone had gone rogue again and the flimsy, nylon thing tied to my bike rack had become a general's tent within a Roman encampment. I told him that 'tent' was probably better, so he typed something else into his phone, looked up and said:

'I sorry. I have a lexical problem.'

But, as always, Liyan had a plan and a proverb too.

'We should stay in guesthouse tonight. Rest. To chop tree quickly, spend twice the time sharpening your axe.'

Given that it's notoriously hard to argue with a proverb, particularly Chinese ones, which are all true, I agreed and we took a day off. We left the guesthouse only once, to pound the ice of some one-street town, where a straggle of vendors were touting pigs' heads flaked with snow. We staggered back with ear muffs, clutching them to our chests, like they were tickets to Hawaii.

'Keep feet warm at night and live long life,' said Liyan. So, side by side, hanging our legs off the bed like toddling brothers, we bathed our feet.

That night I had a nightmare about Liyan. He was standing in the road, looking into the sky, gleeful, shouting 'Steefen it snow!' And then, smack, he was run over by a truck and I was left staring at his spectacles, shattered in the snow.

I woke to a snowscape. Liyan was up already, boiling

Coca-Cola and ginger in the kettle, which infused the room with an unwholesome smell. We set off carefully, onto a road patched in black ice. In Xiangyang, the orange-jacketed army who'd swept the streets with tree-branch brushes were now shovelling snow instead, but the women were still dressed in heels and short skirts, teetering over icy pavements. By evening, a three-quarter moon had turned the snow the colour of tin. There was a little light from Liyan's head torch, dancing over the road ahead. Every now and then he'd shout in English, so that I could understand how shit this was for us. 'Too terrible! Too terrible!'

We'd breached Henan province now and hairpins took us to a ski resort where we found a guesthouse and sat, knees up, around a stove. We were silent for a time, before Liyan said, 'I think holiday in Mongolia not so good idea.' I said nothing, and then we both exploded into laughter.

It was goodbye at last in Luoyang: Liyan and I were destined for different snows. He was up before dawn to cover 500 km in three days, west to his family's village.

'I worried about you,' he said from the door of our hotel room as we said our goodbyes.

'Don't be worried, mate, I'll be fine.'

He left with me a smile and a nod, but I harassed him down the hall with questions.

'Liyan, how do you say rice? Liyan, what time is check-out? Liyan! Please! Show me how to live!'

Alone in China, again. I consoled myself: despite the odds, we *had* been a good team. I was armed with proverbs now. I could ask for a toilet. I packed my panniers, folded up my praetorium, and set off, less fearful of China than before.

*

Thomas Stevens, an Anglo-American, was one of the first people to cross China on a bicycle and, more famously, the first person to cycle around the world. Back then, there was no clear characterisation of this, you simply had to pedal off in one direction and return home from the other, ideally with more facial hair than you left with and some stories about the treachery of the natives. Photos show a strikingly moustached, downright beautiful gentleman, perched on a 50-inch high-wheeler bicycle. He waved goodbye in April of 1884 from San Francisco, packing a Smith and Wesson and a stash of gold and silver, the nineteenth-century debit card.

Stevens might have ended up a Victorian poster boy for machismo and daring, but back then he was unproven – his sallies on a high-wheeler hadn't extended much further than Golden Gate Park. Consequently, he had his share of doomsayers, one playfully remarking that most of the world's roads were unsuitable for such a mission and that he should carry a plank with him, endlessly laying it before his wheels.

Stevens, like me, was 29 years old when he set off around the world. His was the pre-pneumatic tyre age, and he kept mainly to mule tracks, falling off his bicycle with 'undesirable frequency' – which must have effing hurt, considering the altitude of the thing. He reached Hong Kong by steamer from India and, with permission from the viceroy, entered China, a country mind-bogglingly different from that of today, given the intervening upheavals. There were Buddhist shrines and shops selling tiger skins – a tiger hasn't been seen in the wild in China for 25 years. Women hobbled on feet deformed from binding, slaves piggybacked their mistresses, and almost every inn swirled with opium smoke. The roads were horseshoe bends around 'a marvellous field-garden of rice, vegetables, and sugar-cane', the rivers were busy with sampans and junks.

Digging into his account, you sense that Stevens was both

tickled and nettled by the attention he received in China. At an inn, in what was then 'Nam-Hung', a crowd outside his room smashed in a door panel so they could peep at him through the crack. Not satisfied with this, they began to throw turnips at him, in a happy, prankish way, trying to get a rise. But with the language barrier unclimbable (it was illegal in China to teach a foreigner Chinese for much of the nineteenth century), Stevens did what he could to engender trust – by smiling often, handing out peanuts or small coins to children. His mishaps when endeavouring to eat with chopsticks proved an ingratiating hilarity.

Entering larger towns he saw 'people in gorgeous apparel buzz all about me and flit hither and thither like a nest of stirred-up wasps'. In more rural places, people scuttled away or watched anxiously from doorways. Deeper into the country, there was frank aggression. For China, this would be the 'century of humiliation' – war with Japan was around the corner, conflict had sparked with Britain and France, and China had ceded land. 'Fankwai!' the men shouted at Stevens, 'foreign devil!', and he was pursued through one village by a band of men with flaming torches. In 'Kui-kiang':

> My coat-tail is jerked, the bicycle stopped, my helmet knocked off, and other trifling indignities offered; but to these acts I take no exceptions, merely placing my helmet on again when it is knocked off, maintaining a calm serenity of face and demeanor.

There has been a long history of misunderstanding in the thorny relationship between China and the West – excuse the umbrella term. In the eighteenth century, those beyond China's realm were branded barbarians, and it stood to rights that, further out, they were more barbarous. China underestimated

European sophistication and Europeans underestimated the Chinese. Even now, mutual wariness and suspicion persists. I reflected on what China had meant to me before I'd started out. I'd been regularly reminded of human rights abuses and restrictions on journalism. The bitter unfairness of Tibet was prominent in my mind. China was odd, frustrating, lurking. Even on the subject of its economic rise, a smug western media celebrated in the event of a relative slowdown, or wrung hands about what the country's growth meant for everyone else. There were insinuations: China was unstoppable, grotesquely big and busy, dirty and dangerous for the world. Not that such a view was entirely unfounded; it was simply too narrow and tinged with that ancient anxiety about 'people from over *there*'. I'd heard a great deal about China's addiction to coal, but, as I cycled across the country in the shadow of vast wind farms, I could sense the switch to renewables, which will contribute around a quarter of China's power by 2030; a rapid, radical shift in policy that other countries have struggled to match.

As with all travelogues, Stevens's account can speak as much for the country of his birth, in his case, Britain, as those places in which he travels:

> All through China one cannot fail to be impressed with the freedom of inter-course between people of high and low degree; beggars with unwashed faces and disgusting sores and wellnigh naked bodies stand and discuss my appearance and movements with mandarins of high degree, without the least show of resumption on the one hand or condescension on the other.

*

While Stevens ended his Chinese leg in Shanghai, I was well

into the north of China now, having reached the Yellow River, China's other great waterway, which meandered above Zhengzhou. Beyond, I found a land of canyons, frozen rivers and farmed terraces. I pedalled past caves like keyholes, cut into the sandstone.

This was the Loess Plateau, an arid band of the northern silk road, marked by silty soil left behind by ancient windstorms. More than 30 million people are cave dwellers in China and most live here, in Shaanxi. People had occupied these caves – stubbornly warm in winter and cool in summer – for at least two thousand years. This is how the Shaanxi earthquake of 1556 became the deadliest on record: 830,000 people were entombed, 60 per cent of the region's populace wiped out in a trip of the earth's crust. The number of people living in caves today is falling, and many caves lie abandoned as young Chinese have floated off to the cities, leaving me with a choice of bedrooms.

For a day, I padded the tight streets of Pingyao, a walled financial centre during the Qing dynasty and now a tourist stop, full of jinking electric scooters, yapping toy dogs, and Taoist and Confucian temples. Health shops advertised cupping and the overzealous-sounding 'Ear mining'. Restaurant menus boasted of 'Pork elbow' and 'Cow tendons with coriander'. Signs said 'beware of falling objects' and were placed under clear sky, and on top of the walls, which made me wonder whether this was in relation to a specific threat or just general life advice. There were dangers aplenty: 'The senior, the children, the disabled and the pregnant women should have a guardian, to implement guardianship, in order to avoid some sharp-edged situations cause the damage.'

Around the time of the Beijing Olympics in 2008, great efforts were made to correct signs and menus, to delete the likes of 'Government-abused chicken' and 'Grilled Enema'. But

Chinglish, and all mistranslations, are wonderful, not because they allow us to laugh at mistakes, but because of what they reveal about how staggeringly various, unwieldy and evasive language can be. In the tourist toilet, the sign over a urinal was the most joyously befuddling. There were two stick figures, one large and one small, obviously denoting a child and an adult. Underneath the respective figures were the words:

Urine into the pool you short
The urine to the pool you soft

*

There were mileposts on the road to the border, though miles seemed to matter less out here, in the relentless, wan space. This was the Chinese province confusingly known as Inner Mongolia, a derelict world, massacred by an extreme climate; clutches of tumbledown brick houses, stubbled grass, and bony sheep, their fleeces coloured somewhere between the sand and the dove-grey snow that gathered in the dips and gutters of land. I felt Mongolia itself drawing close now, as herders watched me pass by, woolly-hooded, faceless in the howling dust.

12
Lub-dub

A big metal rainbow had been built over the road on the edge of Erenhot near the border with Mongolia. An appeal for hope, perhaps, given the bloody history of this particular international relationship. It was nice and peaceful now though, just a Chinese border guard in a khaki felt coat and fur hat. He considered me for a few moments. 'Bicycle no!' he decided at last, wagging a finger – a common decree at international borders, no point arguing. I heaved my bike and panniers into a grey, clangorous, Soviet van – the sexily branded UAZ-452, which looked prone to breaking down but could probably be repaired with a well-positioned kick. Together with a scrum of vodka-scented strangers, I rattled from an industrial superpower to the fabled sparseness of Mongolia.

Sorry: Mon-go-leee-aaa! Say it with fire. I was primed for a gutsy, untamed, relentless place; six-year-olds hunting with eagles, packs of drooling wolves, heavy men wrestling to the death in snow drifts, that kind of thing. Instead, I got the border town of Zamyn Uud, which at first glance was just a subtler version of its Chinese counterpart. No wrestlers or eagles, just another run of shops. But then that's travel, part wishful thinking, part re-adjusting your view.

I counted four scripts on the shopfronts though: Latin, Cyrillic, Chinese and traditional Mongolian, and at least the shops were stocked differently inside. I could treat myself to seventeen brands of vodka now, or a frozen sheep head with a bloody severed neck, bagged in clear plastic, as if the animal had been suffocated before decapitation. I set off again but didn't get far. A dog, barking murderously, had fought its way under some fencing and I froze as it began bounding towards me.

Here's the thing: Mongolian dogs outsize most Chinese pooches by 20-fold – they've been bred to dismember wolves. I'd been assured that the single most useful phrase in Mongolian is 'Nokhoi Khori!' ('hold your dogs!'). It's what you're supposed to wail as you approach a ger – the round, white-felt nomad's tent (never call it a 'tent') – because more customary phrases, like 'thank you' and 'please' may not be possible, or indeed appropriate, after a lively mauling. Luckily, the creature retreated, for no obvious reason, unless my quavering orders to 'shoo' had actually worked.

I found the main road, which strayed deeper into the Gobi and would strike Mongolia's capital, Ulaanbaatar, in 400 miles or so, but little else before then. Snow patched the desert like land on a globe, peninsulas and archipelagos of hoar frost and ice. I stopped to investigate any glitch in the monotony. The leitmotif was expiration: a rotting vulture, a plundered hatchback. The carcasses of Bactrian camels looked dissected and plasticised, like props from an eighties horror film, pooled in snow like white blood. Vultures had disembowelled them and plucked out their eyeballs, but with hides and hooves intact, there was a grisly semblance of something undead.

That evening I camped on the lee side of a bluff, the sky a royal blue with an inkling of stars. An hour later, I found the haze of Andromeda, our nearest galaxy, and giggled to myself because it reminded me of an Annie Dillard line. She writes about feeling awed as she contemplates the contrasting scale of things, and then threatens to take an amoeba outside to show it Andromeda and 'blow its little ectoplasm'.

I thought about how cosmic the human body can sound now, as I admired that spangled sky. Matter exists in our craniums and in the universe. Our awe for the celestial finds expression in our awesome anatomy: there is a stellate ganglion in the neck, a semi-lunar valve in the heart.

I woke to the wind, a soft, scratchy sound, like the start of a record. There was a sense of anticipation too, though the hours came and went, eventless. Jets trespassed across the blue sky from time to time, scarring it with contrails, not diminishing the wildness, but lending perspective, as if the sky and steppe had equal dimensions.

In the afternoon I stopped by a few houses by the railway line in order to resupply with water. A tall, vigorous man in a deel emerged to greet me. I mimed a drink, but he laughed and hurried me inside his home where there was a feast – a tower of biscuits, a block of meat, gristle and innards. The man cut me strips of meat and poured out some *airag* – fermented mare's milk. I don't want to be disparaging about a drink that commands such national pride, so I'll just say that it complements cold, ambiguous carrion. I tried some *aaruul* next – named, coincidentally, like the sound you make as you ingest it – a hard, emphatic cheese, or to put it less benignly 'National Dried Curds'. Mmm.

The wind had been a useful ally all morning, rushing at my heels, but now it turned face, and pierced me with cold. Midday now, and the temperature was still minus 15 degrees Celsius. I hoisted up the hood of my down jacket, yanked on gloves and over-gloves, and covered my face with a neck gaiter, the wind shrink-wrapping me in my clothes. It's an unpleasant sensation, cold eyeballs. Ice crusted my beard and every appendage went numb (*every* one). Ahead of the Mongolian winter, I'd been advised to put a sock down my trousers but, horribly conscious of what might happen if I didn't, I'd put down three, making myself a ridiculous woolly condom.

Silence had been a solace in a world so howling with cars, but now the wind's racket was overbearing too, and I longed for something to displace it, even the sound of engines. And at last they did, as I neared Ulaanbaatar. I noticed the city's pall of pollution first, an aloof, hovering greyness, then gers and wooden homes, an expanding community of one-time nomads, growing by forty families a day, as higgledy-piggledy as a farmyard.

Ulaanbaatar contains around half of all Mongolians and most of them were adding to the smog, much of which came from coal-burning stoves. In an effort to combat the problem, the government had begun replacing the stoves with more fuel-efficient models, which meant that canny Mongolians could burn even cheaper coal, keeping the air pollution at an even smogginess and the respiratory disease at a steady mortality. When they tested the air over the ger district, they found its quality to be worse than the infamously airless atmosphere of Delhi.

Mongolians still contribute much less waste than an average westerner of course, and pollution is disproportionately caused by richer people and visited on poorer ones. But at the sight of so much smog, I was reminded that periodically, and for centuries, humankind as a whole has been compared to a plague, parasite or cancer of the planet. As populations surge, as the climate crisis ramps up and the world grows feverish, there is again traction to the idea that we – human beings – are pathological. Malignant spread is rapid, uncontrolled, homogenising, destructive and colonising, after all.

I don't buy it. A tumour or plague cannot communicate and cooperate like we do, or make decisions to improve and self-regulate. There are other, kinder ways of looking at us.

The physician and essayist Lewis Thomas wondered if we could be, all of us, more like one organism, as cooperative as ants or swarming insects or schools of fish or flocking birds, functionally at least, a great multi-organism organism. Or perhaps, he suggests, if we are just part of an organism, we could operate like the planet's nervous system, a planet that is in fact a loosely formed creature, where individual human beings are nothing more than 'a cast-off cell, marooned from the surface of your skin'. The Portuguese man o' war comes to mind. Not a jellyfish, but a siphonophore, it's a colony of many beings cooperating, so it functions as a single organism. It's made up of polypoid zooids: some feed, some catch prey, some reproduce, and so on. They share a kind of nervous and digestive system, but each is an organism and none can exist on their own. So who the hell's in charge?! Spend too long lingering on the world being like a Portuguese man o' war and you'll have an existential meltdown.

This earth-as-organism metaphor might be a bit New Age, a bit teleological, for most of us, even if you see something like disrupted homeostasis, that stable equilibrium reached by harmonising physiological forces, in our current imbalances: the climate change, conflict and over-consumption. But perhaps it's not entirely fanciful. As well as the entire planet, cities have been compared to organisms too, in both classical studies and in urban sociology. They both consume energy and resources; they produce information and waste. They can be highly complex and they require close integration and networks: roads, railways, electrical lines; vascular systems, lymphatics and neurones. Despite the incredible diversity in the sizes of mammals, they are mostly scaled versions of each other, and across the entire spectrum of life, most physiological variables scale with size in a remarkably simple and predictable way. This holds for how long they live, to how many hours they sleep, to the famous observation that every living being on earth gets about a billion heartbeats worth of lifespan. Larger organisms live longer, but their hearts beat more slowly. And when you learn that cities, like organisms, also exhibit this universal law of scaling too, you begin to wonder about the nature of life, and about what might lie beyond our world.

*

If your aspiration as a traveller is to blend in, Richard Burton-like, you might do better than travelling on a bicycle. I lumbered into Ulaanbaatar with overflowing panniers and saucepans stuffed under bungees, looking like 'an aftermath of a gas explosion', according to one description of a touring biker I'd once read. It's even harder to hide if you're cycling around the coldest capital city on earth in the dead of winter. But this conspicuousness can sometimes work in your favour too, and as I headed towards the city centre, a young Frenchman called Frankie chased me down on his own bicycle.

'Hey, where are you staying? Come and stay with the Dutch guy!'

I set off with Frankie to find 'the Dutch guy', wondering if I was about to be introduced to the kingpin controlling Mongolia's narcotics trade. He'd be easy to find, probably, as the city seemed half-empty, and even Chinggis Square outside parliament had a derelict feel. A stony Genghis Khan (Chinggis, if you're Mongolian) looked out over a flight of steps, confidently tyrannical, his barren paving stones crossed occasionally by solitary men, moving fast against the cold, and once, a trio of girls, encaved in furry hoods.

I followed Frankie, past the State Department Store, a nail shop named 'Destroy' and 'The first Irish pub in Mongolia'. Past unpeopled cafés called Los Angeles and California. Old buses, retired long ago from the streets of Moscow, rattled along. Statues of Stalin and Lenin had been ripped down years ago, while Chinggis Khan's personality cult was approaching peaks last seen in the thirteenth century. Frankie pointed out a brickwork set with bronze reliefs of the Beatles. It commemorated an old infatuation, when vinyl was smuggled into the isolated, communist Mongolia of the sixties and seventies, when kids gathered to sing 'Let it be' from the stairwells of tower blocks. On the other side of the brickwork, there was another relief: a lonely, long-haired Mongolian man strumming his guitar. He is isolated from the Fab Four, just as Mongolia is apart from the world.

Eventually we arrived at a small home. Frankie knocked and the door was thrown open by a man in an aqua shirt, embroidered with gold. Froit had a grey moustache and long grey hair, swept backwards. He was Dutchly tall.

'Frankie, you brought me a cyclist! Come in, come in. I have a warm house, I want to keep it that way.'

Froit had clearly been in Mongolia for some time – it would take years for a life to become so styled with knick-knacks. There was a glass case with miniature galleons in bottles, Buddha, a horsehead fiddle, rugs woven with the images of bearded old men, looking divine.

'It's a good time to visit!' Froit said. 'It's New Year soon. That's when Mongolians get together and discuss who's left who, whose house has burnt down, that kind of thing.'

Froit had drifted to Mongolia in the 1970s with a clique of European artists and 'anarchistic street performers', all of them taken by the adventure of Mongolia. In the capital, they were commissioned to create a large *ulzii*, or 'endless knot', a sacred mandala linked with Buddhism. After Froit's first wife suffered heart failure, he remarried a Mongolian woman and then a neighbour taught him how to build a ger. A hands-on sort, Froit had found his niche. When he talked of gers, his passion boiled over.

'You have to build them *by the millimetre*. A tepee has 17 sticks, ha! A ger has 81!'

Armed with this new craft, Froit went into business constructing and shipping fireproof gers for use in festivals across Europe, bringing a slice of the steppe, and none of the hardships, to Glastonbury. Unfortunately for Froit, there had been a fire recently in his house, and many of his fireproof gers had burnt down.

Over coffee the next morning, Froit asked if I might drop by and see his wife's sister, who had hepatitis and was growing unwell. I worried about what magic they might expect me to achieve, but Froit and his wife appeared a little emotional when they talked of her, so the next morning we left, driving past Chinggis Square, where heavy, fur-hatted guards were standing on the steps of parliament scrutinising a clutch of protesters. The cause célèbre was a foreign-funded gold mine inside a national park near a sacred mountain, palms no doubt greased in the deal-breaking. Just beyond the square, I noticed a conspicuous foreigner loping down the street, a black man, six foot eight, perhaps. Later, someone cleared up the mystery: 'Any Mongolian politician worth their salt owns a horse, a TV station and a basketball team.'

*

We stopped outside a wooden two-storey house. Hulan, a lady in her forties, was upstairs in bed, with what doctors sometimes call an end-of-the-bed diagnosis, because a clinician should be able to make one from a distance. The whites of her eyes, the sclera, were tinted mustard yellow. Her abdomen domed the sheets. Jaundice, coupled with such distension, boded badly – it almost certainly meant that her peritoneal cavity (the space in the abdomen between the bowel and abdominal wall) was full of fluid, a condition known as ascites.*

It's incredible to consider that most people don't really know what the liver does. It's the second-largest organ in the body (after the skin, which you'll know from pub quizzes) and people rarely know much more than that we can break it by boozing. The fact is, the liver does lots of complicated stuff to keep you going. It's a kind of biological larder, disinfectant, factory and food processor all in one (and more besides), so liver disease leaves a remarkable range of marks on the human body, related to what, as it fails, it becomes unable to produce or break down. A diagram of these stigmata in textbooks is a mayhem of lines and arrows skewering every body part: from the skin to the testicles. Men with liver failure grow breast tissue as precursors to oestrogen build-up. People swell, ammonia dulls the brain and causes a tremor of the hands, like the slow flapping of bird wings. The breath gives off a sweet and sickly smell called *liver foetor*. Small dendritic blood vessels – *spider naevi* – break out on the skin. The palms of the hands can redden, the parotid glands swell, rounding the face. The fingernails can turn white and bow outwards. It's not hard to tell a patient with chronic liver disease.

I knelt to examine Hulan. Every medical exam begins like this, with a moment of stillness, for inspection. Veins meandered over her belly. I placed my hand over them and tapped on my finger near the umbilicus: the note was hollow from air in the bowel loops. I began tapping laterally and moving

* The other causes of abdominal distension are made easy to recall for first-year medical students in the six Fs: fluid, faeces, fat, flatus (farts, to you and me), foetus, fucking big tumour.

downwards towards her flank until a dull, less satisfying note sounded back. When she rolled onto her side, the note became resonant again. This 'shifting dullness' suggested that the distension was caused by fluid, which can swash this way and that, dependent as it is on gravity, unlike the other Fs.

In medical parlance, Hulan had end-stage disease, but you can call it dying. Complications might hurry it up; she was, for example, at risk of cancer, of infection – since the liver produces antibodies – and of internal bleeding, as the liver produces clotting factors. I'd witnessed this occur catastrophically once, from engorged veins in the oesophagus called varices. It was my first year as a doctor, and I was on call in the witching hours of a small regional hospital. A nurse asked me to see a patient complaining of abdominal pain. I sat down beside him, a burly one-time prop forward whose penchant for after-game whiskeys had decided his rugby career. He looked sweaty and troubled. With a groan he slumped forward, retched and vomited into a cardboard bowl. When he leant back, I saw that it contained nearly a litre of fresh blood.

I called for help. Nurses snatched at the curtain. I crouched by his side and threaded a cannula into his arm, while someone raced off to call the emergency medical team. Now more blood was escaping; it poured fast and black from his bowels as his blood pressure tumbled, his skin blanched and his consciousness waned. The crash team arrived, we scrambled to resuscitate, pouring in blood transfusions and fluid, but it was a colossal bleed and he went into cardiac arrest before the surgeons or gastroenterologists could intervene.

For liver disease, Mongolia had been the perfect storm. Vodka was involved for some, especially men, and it was rarely in short supply. Relative to population, Mongolia had more shops selling alcohol than anywhere in the world, and post-independence food rations actually included bottles of vodka.

Viral hepatitis was an epidemic too. During Mongolia's days as a Soviet satellite there was no syringe factory in the country. For years, medics reused equipment in surgery, dental procedures and childbirth. Hepatitis C raced through the populace until Mongolia had the second-highest rate in the world and 10 to 20 per cent of adults had been infected. Today, cirrhosis or liver cancer causes 15 per cent of all deaths in Mongolia.

In the early 1990s, when the Soviet Union sundered into new nations, the Asian Development Bank and the IMF applied pressure to Mongolia through the benignly named process (it was anything but) of 'structural adjustment', which has since been widely criticised for devastating health and education systems. In this neo-colonial exchange, Mongolia got loans but was compelled to introduce user fees for medical care and to slash access to it. For the next decade, healthcare across the country decayed. The ratio of doctors to patients plummeted. Doctors went unpaid and scarpered abroad. Hospitals ran out of surgical thread, and ambulances, vital in a country scattered with patients, fell into disrepair. When the country's only dialysis machine in Ulaanbaatar broke down, six patients died. The Mongolian elite rushed to neighbouring countries for medical treatment but the rest had to pay for their own bandages or resort to quackery, which was booming. The traditional remedies for liver disease included eating the liver of a wild boar and drinking a broth made from the hair of rutting camels, sold in match boxes. Along with hepatitis, TB, cancer, heart disease and vitamin deficiencies all resurged.

Hulan was not bleeding, nor septic, as far as I could tell. She was already taking drugs called diuretics that remove salt and fluid from the body. She was due to attend hospital the next day for drainage of the fluid in her abdomen and for blood tests. I could do little more than clean and dress the wound made by a previous chest drain. Perhaps she would have benefited from a transplant at some point, but few livers had been transplanted in Mongolia, and her family couldn't afford to follow the richer ones to Korea or India.

I could do nothing else to help, but Hulan and her family thanked me generously and I recalled their appreciation, their grappling for small victories, when, a couple of weeks later, Froit told me that Hulan had died at home. I could have pointed out this possibility of course. I may have explored her expectations had I known more about her illness, her test results. I may have 'doomed and gloomed', as a consultant I knew used to say. Doctors are not always good at this, and I sometimes have to remind myself that compassion is not simply keeping the patient happy. There's always a temptation to rose-tint the future, or, as I had done, simply not mention it at all.

*

At last it was time to go, headlong into the steppe. After a few kilometres there was no hint that a world capital collected close by – just shivering grass and salt-and-pepper hills. Chinggis Khan is rumoured to have been buried somewhere in the great expanse of Mongolian steppe, with 40 young women and 40 horses. Nobody knows where. I felt endangered by the vacancy ahead, chiefly by its vastness and cold and lack of concealment. Rumour had it that robbery was not uncommon in the west. This was coupled to an absurd but somehow likable notion that if a thief has the savvy to steal your stuff then they deserve it.

Almost all the paved roads in Mongolia converge on the capital, making a kind of red spider on a map. Further out, the road surface is determined by the season, varying from sand to mud to snow and ice. An unpaved Mongolian 'road' beyond the capital is in fact a network of rough, transient trails. Cars forge new tracks continually, and without knowing where to put signposts, no one goes to the trouble. To add to the confusion, some Mongolian towns have identical names and the duplicates can be painfully close to one another. My map told of Altay near Altay. Tosontsengel wasn't far from Tosontsengel. Darvi nudged up against Darvi.

About 100 km north-west of Ulaanbaatar, the pavement ran out and then the dirt track split four ways. There is a phase of getting lost, the earliest phase, when you're absent-mindedly not quite where you thought you were, which can be quite pleasant if you can relax into it. This is never possible in a Mongolian winter, where a wrong turn can mean horrible cold and something terrible happening to your knackers.

From a ger up ahead a man emerged and I sought directions. He didn't point me down any of the nearby tracks, but to a point on the horizon instead. Another quirk of Mongolia: I was often pointed through mountain ranges or lakes, and even in towns Mongolians sent me through tower blocks, as if they had X-ray vision, or simply imagined the land in a more natural, uninterrupted state.

Before I set off again, the man invited me inside his ger and I accepted. A woman was butchering a lamb using a stone axe, and soon we were ripping meat from bones. My host had two sons, teenagers, one of whom wore a thick blue deel and had apple-red cheeks and big, roughened hands – he could have been an extra in a historical reconstruction of the Mongol empire. The other was a student, home from the city, who fooled with his haircut and strutted about in jeans, as if ready to meet friends at a multiplex cinema.

'Mal sureg targan tavtai yuu?' I tried, one of the few Mongolian phrases I'd memorised, which means 'Are your animals fattening up nicely?' The man heaved backwards and with a slow proud nod, confirmed that yes, his goats were fat and hardy. Or, he'd listened politely to a stranger making sounds that bore no semblance to those of his language, and it was easier to nod than to ask what he was on about. As his wife moved from her stool towards the fire, she farted and let out a nervous laugh. The man looked embarrassed for a second but then laughed too, and that set me off. It's a fond memory: sitting in a ger with nomadic herders, somewhere in the vastness of frozen steppe, laughing about a fart. Disband the UN, we've got this figured out.

I set off again. The steppe was livelier now, with horses cantering towards me, lurching away, the wind flaying their blond manes. Vultures sat bulkily, surveying waddling sandgrouse like patient police officers. A low, brown smudge was the first hint of the taiga, an immense boreal forest that almost wreathes the planet, sheltering wolves, moose and wild sheep.

I was excited to reach the village of Khatgal because a lake called Khövsgöl stretched north of the village, almost to Siberia. For some time, I'd been possessed by the idea of cycling across a frozen lake, and in expectation, I was packing a pair of spiked ice tyres to grip onto the thick skin of ice. Khövsgöl was 260 metres deep in places, and between December and June, almost three-quarters of the freshwater in Mongolia was held here, arriving in 96 rivers, while only one, the Egiin Gol, left the lake. I yearned to stand on top of such water, to feel what I imagined to be a vertiginous thrill.

I went to the shore for a recce, padded carefully onto the ice, lay down, and peered through a hatchwork of silvered fissures, like reams of tin foil suspended in glass. There were tiny spheres, bubbles and snow, minuscule ice-worlds. Amid the cracks and snaps came other sounds, unexpected bursts, and low thumps, like heartbeats, as water bubbled up through fresh rifts in the frozen underbelly, *lub-dub, lub-dub, lub-dub*. I sensed my own heartbeat too, though I needn't have worried. The ice, cast by a Mongolian winter, was over a metre thick in places, thick enough to hold me, certainly, thick enough to hold a truck too, or most of them anyway. For years trucks had plied Lake Khövsgöl in the winter, delivering oil to Siberia. At least forty had crashed through the ice and sunk to the lake floor. In the early 1990s – at around the same time that Khövsgöl was incorporated into a national park – authorities banned heavy vehicles in order to save further pollution.

I left the lake and found a guesthouse where I met a local guide, the sturdy Ganbat. He was hunched over a rumpled map, 1 cm per square kilometre, once classified, a relic of Mongolia's Soviet years. Ganbat pointed absently to the spot where, two years ago, he'd lifted a bunch of Russians from the roof of their car as it tipped underwater on sinking fragments of ice. 'City types,' he said, with a smile. 'They don't know the weak points. The ice can get thin around these spits of land, and here, where the rivers come in. And anywhere the fishermen have cut holes.'

Alongside the fishing communities there were reindeer herders too, mostly in the Darkhan depression, a waterless mirror of Khövsgöl to the west. There was even the odd foreigner.

'An old German lived here for a few years,' said Ganbat. 'People called him Grey Wolf. He put up an advert in Khatgal asking for a Mongolian wife, but none of the single women were interested.' He laughed boyishly.

'It's not good for a man to live on his own. They all go bad this way, better to die than live without a woman. One day there was no smoke coming from his chimney – some of us went over, found him dead. He'd written instructions, wanted a sky burial. We carried him into the mountains, left him for the animals to eat. Couple of men from the German embassy came by, asking questions.'

He shrugged.

'We were only doing what he wanted.'

I wondered if Grey Wolf had known that he was dying when he came to live beside a remote ice lake. Perhaps he saw a futility in detaching loneliness from death, wherever it occurs, or perhaps loneliness was a small price to pay for the privacy and beauty of the land.

There were a scattering of other foreigners here too, mainly missionaries, from the US and Korea. Ganbat had little time for them. 'If you're poor, uneducated, have problems, maybe you go to their church. I don't need Jesus.'

I'd heard stories of missionaries venturing into the steppe, giving out bibles. Nomads accepted them willingly, ripped out the pages, rolled them up, packed them with tobacco and smoked them as cigarettes. Ganbat had his own tale.

'A few months ago, I helped this old American guy, a missionary. He'd lost a car tyre somewhere out near the lake shore. I searched for a long time and found it, hauled it into my car and brought it to him. He didn't thank me, just kept thanking the Lord and saying that God had found it.'

Ganbat huffed and shook his head.

'It wasn't God. It was me.'

*

I slept that night in a ger outside the guesthouse, or tried to. I was sharing it with a former wrestler, a vast man with cauliflower ears. Sleep talking is always a freaky business, but when it comes from a giant, somewhere miles from rescue, and in Mongolian, trust me, it is bone-chilling. Mongolian, by the way, sounds like someone is waterboarding a Klingon.

I returned to the lake the next morning, this time with my bike, rolling it onto ice stippled with grainy snow. Carefully, I saddled up and pedalled through a harbour, skirting two ageing ships anchored by the ice. Distracted, I felt the unbidden slip of my front wheel and went down on my shoulder, hard, wincing with the premonition of a crack and a bitter engulfing. The ice felt like titanium.

By evening the ice croaked and muttered even more. Sometimes it snapped so loudly it echoed off the mountains and could be felt as a convulsion under my feet. At dusk, the sky turned lavender, the ice, stone-black. I pitched my tent on the ice, a hundred metres or so from the shore, stood back to admire its woozy reflection and then settled in for a cold night.

The sun rose from a gold band, seaming the taiga to the sky. It was the coldest night I have known. The temperature had fallen to minus 38 degrees Celsius, and yet, somehow, I'd slept well on the ice. Perhaps I'd hardened to the cold; certainly, I was better prepared (three sleeping bags, two inflatable sleeping mats). I sensed a yawning distance between me now and the man who had shivered pathetically through France five years ago, cursing his fingers and the universe. Thinking ahead though, I had more than the cold to worry about. For a thousand miles, heading west, I expected only a few run-down towns and villages and barely any shops. 'Bakui', they'd say, which seemed at times the national motto, meaning 'isn't' or 'aren't', as in 'we don't have it'. I'd push on, it was habitual, but I was beginning to wonder if pig-headedness, a precondition for adventures, is a type of stupidity too. Over the last few years I had become expert at muting any kind of thoughtful, internal dialogue. I'd resisted the temptation to question what it was all for, what the benefit might be, what else I could be doing, and whether it was foolish to cling to old ambitions in spite of new ones. 'Be naive enough to start, stubborn enough to finish', they say, a mantra I'd taken to heart. Maybe I *was* burning out, but there was still a powerful sense that I needed to finish what I'd started. But why? For what? (Shut up, SHUT UP!)

The steppe was streaming with cloud shadow on the day I left Khatgal, and the air felt only cold, not glacial. Had winter finished with me yet? Being on the cusp of spring was an almost elating thought. That afternoon I watched brown water prattle through a narrow channel in the ice and melt an adjoining stream, and it seemed like I was watching the new season unfold in front of me.

I didn't know it yet, but there is not much that is spring-like about spring in Mongolia. It's not exactly daffodils and stretched-out days and children skipping in parks. There are lambs but small hiccups in the weather can decimate animals and they have to be plucked from the snow by herders and nurtured inside gers. In fact, for most Mongolians, spring is the least favoured season; it's when governments tend to resign and workers protest the most. Even winter is preferred, because April and May can bring savage dust storms from the west.

The lesson began the next day. Dark clouds were tanking over the horizon; a stiff wind kicked up, like a prelude to war. Unless I could pile up dead cows to use as a windbreak, there was no other shelter from the gale. I managed to get my tent up fast, but it was blasted madly, and all night I worried I'd hear the crack of a pole and wake up wearing the flysheet like a maternity gown.

I woke to birdsong instead. The morning was more sedate than I'd seen in months, and the nylon of my tent had tautened to hot silk. If gales can be haunting, there is also something uncanny about the steppe when the wind dies down entirely, and I shuddered as I looked over the tiny town of Tsagaan Uul two hours later. I felt like a medieval knight advancing on a fortress – the town was fenced off against the wind and I must have been visible to the townspeople for miles, creeping lonely towards them. Shaggy dogs stalked the fringes of the village and two horses fought viciously in the street, biting, raging, kicking up dust. I moved on.

In the afternoon, I passed a single, deserted house, represented by a black block on my map, an absurdity, given the scale of one in two million. Uliastay was next though, and I found a guesthouse between a sign for 'World Vision' and a pile of empty vodka bottles. The owner, showing me to my room, hacked up phlegm, gobbed in my toilet and handed me the keys. Sounds of fucking came through the walls, then the vibrations. I had to get out of here.

As an immediate plan, however, it wasn't a good one. The forecast suggested that 15 cm of snow would fall, so I took the next day off too and watched a blizzard whirl in, harangue and layer the town. The following afternoon, the sky contouring the Altay Mountains was clean and blue so I set out for a pass marked by a sacred heap of stones, an *ovoo*, pushing, mostly, the snow too deep to ride. The view at the top was gut-wrenching – mountains stretched away, snow-thick and endless, the trail just a light scratch across the land, sutured in snow. For all I knew, it was the same story for one hundred kilometres or more. I didn't have enough food or stove gas for such a battle, and with my left knee beginning to ache, frustration was verging on fear.

After a few hours, with plenty more pushing, a string of pylons came into view and I sent a little scream of joy into the wind. The trail was hardly visible anyway, so I went off-road and followed the pylons into Altay. My knee was clearly swollen now. Patellofemoral syndrome? Patellar tendonitis? Fucked knee syndrome? Whatever, some sort of overuse injury, probably, which was frustrating because treatment for overuse injuries is to stop using the bit being overused, and my visa was about to expire and there was no traffic to hitch a ride. I rested for a day necking ibuprofen, with a pack of ice cream on my knee, though I kept eating the ice cream and it hurt to get more so I just sat glumly and watched my knee swell instead.

After a few days the swelling eased enough to risk cycling again. Soon, mountains, many-toned with mineral ores, went drifting by in the distance. In the afternoon a motorbike stopped up ahead: a man and a boy were waiting for me in the road. Generally, I'd been greeted with a smile all over Mongolia, but this man was poker-faced, his right eye red and infected. The boy looked no more than ten years old, and had bruises over his forehead and cheeks. The man stepped into the road, and as I stopped, he made a slitting action across his neck and pointed at me. He was either offering to trim my beard, or he was threatening to violently end my life.

Luckily, my policy for both scenarios is the same. I gave him a whopping grin, like someone who doesn't understand. I offered my hand too but he didn't take it, reaching instead inside his purple deel. I braced for a weapon, but a bottle of vodka emerged, which he tipped to his lips, and in that long glug, the bruises on the boy's face made a miserable sense. I would do anything to avoid a fight, especially out here with no hope of intervention, but as I moved past him, I knew that it might not be my choice. No turning back either, if that's the way it went. I felt my whole body prime, my fist tighten, and each heart beat was a blow to my chest. He didn't move though – maybe he noticed my readiness – and I set off again, fast, tingling with a tide of adrenaline.

I spent my last night in Mongolia high on a ridge, trucks chugging to the Chinese border in a trail of headlights far below. It was too visible as a camping spot, and blasted by wind, but it was my final night in Mongolia, and I fancied being closer to the stars. I slept outside my tent, watching meteors ghost across the sky.

13

Revival

By now, there were signs that I had grown far too accustomed to life on the road. Plenty, if I cared to notice them. Whole billboards, flashing by to the left and right, *Burnout*, *Burnout*. When I lost my spoon, I ate with a bicycle multi-tool for four days, and it seemed like this might go on for a while. I replaced hardly anything now. My clothes had been repaired with long, meandering lines of stitches, bringing to mind Frankenstein's monster. I even dreamt of cycling. One night, sleep carried me to a painfully beautiful desert, a flashback to the Namib perhaps, and I took the dunes like BMX ramps. But I woke up in Xinjiang.

Nothing so sublime here, a sky smeared in greys, and lumpy sand on both sides of the road. Yesterday, I'd been relieved to reach the Chinese border, the regimented slog of Mongolia finally over. But this was a rambling, atrophied world, where even the sight of a fence brought relief. Hardly a revolution in fortunes. I'd simply transferred from one edge to another, and the badlands are still badlands, be they Mongolian or Chinese.

It would have been even more tedious had my knee not throbbed with pain each time I pedalled. It was swollen and it crunched sometimes too, a sound like Velcro pulling apart. The medical term for such a nuisance is *crepitus*,* a warning sign of

---

* Also: great name for a posh kid, just imagine: 'Crepitus! Tarquin! Time for supper!'

inflammation. My thoughts drifted to a Welsh patient of mine who'd assured me that he knew farmers in the valleys who used the mechanical lubricant WD40 on their arthritic knees. I've always hoped that was true.

I needed a base camp, somewhere to rest and ice my knee. Fortunately, an Australian cyclist, Sam, had offered to let me recover in his flat when I got to Urumqi, capital of the Xinjiang region. Whingeing and self-pity were not part of the deal, but he'd have to live with that. As it turned out, this was easy for Sam, a cheerful yogi who'd paused his own bike ride across Asia to earn money by teaching English to college students.

I went hobbling about town the next day. Urumqi felt like a city on edge; there was an overtone of something simmering, some balance upset. For more than a decade, money had flowed into this flank of China, Han workers too, their number in Urumqi now roughly equal to that of the indigenous ethnic group, Muslims called the Uighurs. In Urumqi, I wandered past armed police in cages. Petrol stations were cordoned off behind X-shaped fences and there were metal detectors outside shopping centres and parks. The city was not quite the murals on walls, where ethnic people danced about playfully, jubilant in their national inclusion.

China was in a bind. Permeability was a vital condition of the Belt and Road Initiative – touted as a sort of New Silk Road – but while goods moved to and fro, the government fretted about less welcome imports, especially arms and ideology from troubled pockets of other states. This echoed an older feat of the Silk Road, that intercontinental conveyer belt of religions and crossbows. Xinjiang is a large state that shares a border with eight countries, and there are Uighur separatists who would have the -stans start further east, though the idea that China would ever cede land, that an 'East Turkestan' would be born, felt far-fetched. This seemed especially so at the crossroads of Renmin Lu and Heping Nanlu, where a gun-toting soldier stood hip-deep in an armoured car, scanning pedestrians with the sangfroid of a crane watching a river for fish.

On the far side of the crossroads, China seemed to split and a new personality emerged. Now women in headscarves shuffled past the mosque, greeting one another with salaams, and men sat in loose knots about sacks of dried grapes. There was a barrio-like bustle. Even the clocks were two hours apart. Many Han were afraid to come to this side of the city and most had no cause to. The kids were more roguish here, they chased and called after me. Girls wanted a photo. Nobody seemed to be heeding the collectivist mantras of Chinese classrooms, like 'The bird with the longest neck gets its head chopped off'.

Tensions tended to escalate with the temperature. It was now early May and a year ago, almost to the day, two SUVs were driven into a market near here. Uighur militants flung explosives from the windows and the vehicles collided creating a fire ball that mushroomed above surrounding buildings. That, and the 43 bodies strewn in the street, made it an act of violence too flagrant for the Chinese censors to hide and one that the US were conspicuously slow to diagnose as terrorism, preferring at first 'ethnic violence'. One Chinese commentator suggested that if a hail of grenades wasn't terrorism, then 9/11 was a 'traffic-related incident'.

Pinpointing the 'Uighur look' was a mug's game – the streets bustled with Mongol cheeks, Turkic noses, jade and hazel eyes, mousy, reddish and raven hair, a physiognomy the Chinese government had captured with facial recognition technology and paired with voice recordings, leading to accusations that Xinjiang was now a dystopian police state. It was a war against Islam and against any identity that presumed to trump Chinese. Wary of divided loyalties, long beards and face veils had been banned. Down a side street, I found posters reinforcing these rules, with photos of despondent Uighur men under arrest. Looking up, I found three CCTV cameras reading me. Urumqi had tens of thousands more. With the help of AI, human anatomy was being read, segregated and weaponised.

*

With my knee less fat and crunchy, I left Urumqi on a highway lined with barbed wire, broken only at culverts, where I could hide away and catch some sleep. Only a sliver of China remained on my map, but it was still four days before I saw the borderlands in the guise of tall, brumous mountains: the Tian Shan. I ascended, the world green at last, spruce speckling grassy hills. After the blunted hues of the desert and steppe, it was like getting my sight back.

I took a day off, and camped, with Nabokov's short stories, beside the tulips of Sayram Lake, where I took painful plunges, the water a silken sting against my legs, too cold to get used to. Beyond was a smooth highway, just a year or so old, a brave, trumpeting artery of the New Silk Road, not to mention calm management of an epic obstacle course. Such keenly rippling mountains had tested the Mongol hordes for centuries; the modern Chinese made the conquest look easy. The highway was suspended loftily over the valley on limbs of concrete. I squeezed onto it by a break in the crash barrier and let my wheels roll, as audacious Chinese engineers threw me down the mountain on a sequence of bridges and tunnels. I glanced to my right, catching the valley in perfect cross-section, but then I was fired around another bend and all went green and vague with streaming hills.

I crossed the border into Kazakhstan at Khorgas and stopped at a stall selling apples. I'd read that apples had originated from Kazakhstan, in the forests of the Tian Shan in fact, a detail discovered after sequencing the fruit's genome.

I pointed to some small green ones. 'NYET NYET!' thundered the vendor.

'These Jackie Chan apples. You want ...'

He reached down and rummaged in a sack by his feet.

'Steven Seagal apples!'

And with a flourish he held aloft a weighty red example. Over the next few weeks, when I introduced myself as Stephen, it was always Seagal that people yelled back at me – sometimes adding a karate chop – never Spielberg, Jobs, King or Gerrard. Some Stevens have more cachet than others in different parts of the world, especially if they hurt people on TV. In Egypt, my most distinguished namesake had been WWE wrestler Stone Cold Steve Austin.

I'd arrived in Kazakhstan on National Public Drag Racing Day, unless it wasn't and for some mad reason Kazakhs drove like this all the time. Speeding and tailgating were in vogue. One Lada was dragging its fender and scraping the road. Haunted by the vision of bike verses truck, a scenario I'd successfully avoided now for over 70,000 km, I deployed the 'safety shuffle' – a little wiggle of the handlebars – when I sensed a car was approaching too fast and too close from behind. It's a feint, really, to make the approaching driver think that you're ham-fisted and dangerous, which in turn grants a little extra space, provided you time it right. It worked a treat but was so frequently necessary that it seemed to add miles to my days.

I didn't mind the hold-up so much. Rocks, the colour of candle wax, stacked up beside me and the air was flavoured by camomile and sage. It was a Sunday and families were on excursions from Almaty to enjoy these frontier highlands. Cars stopped, windows dropped, and bundles of leftover salads and meat were pushed towards me. I could read Cyrillic and chat a little in Russian, which I'd studied for a few years, and felt glad that the maelstrom of Chinese was behind me.

I woke the next morning to a druggy light and a fuzziness in the northern sky. At first, I took it for rain, but soon a squall of dust was needling my eyes. Trees, corralling the odd farmhouse, were savaged into yogic poses by the gale. The light turned a dangerous cerise, and power cables whipped about like fishing line. Boughs of trees were sheared off and twisted shards of wood junked the road. A rust bucket of a car packed with teenage boys drove beside me, the driver blasting his horn, hollering and glugging from a bottle of vodka. 'HITLER!' he screamed, pounding his fist into his other hand, and I would have liked at least one on the steering wheel. Of all the things to hear a hostile, drunk neo-Nazi in command of a vehicle scream, this ranks as one of the least idyllic. I fixed my gaze straight ahead and after a few minutes they grew tired of me and lurched away.

Further down the road, two old men in suits were stumbling along, and it was impressive that anyone could be so unmistakably drunk from a hundred metres away. Closer up, I saw a great array of military medals pinned to their lapels. They homed in on me like happy zombies and kept grabbing my hand, trying to pull me off my bike, but in a fun sort of way. After a quick chat, I realised why all the booze and fun and shaming (not celebration) of fascists. It was 9 May: Victory Day for Kazakhstan and a bunch of other ex-Soviet republics; the day when, in 1945, Germany surrendered in ink.

Heading west towards Almaty, I passed a lonely shop with a rusty sign that screeched in the wind. I shrunk into a wicker chair on the porch and watched the storm, clawing at some toffee popcorn. Men with pink eyes ambled in and out of the shop, ponging of vodka, sizing me up. The couple who owned the shop and attached restaurant had three kids who looked like they'd caught some of the crashing chaos of the sky. They zipped around in Brownian motion, rebounding off each other and stamping in puddles. Their mother readied borsch and milky tea.

'You should stay with us,' she said, in Russian. 'It's too cold outside. We have a spare room.'

It was at the back of the house and there was just enough space for a double bed and a sofa. I stretched out on the bed and began to read, but a few pages later, the door cracked open. It was the owner. He looked different now, his arms were limp by his side, his stance apologetic, perhaps. The reasons were beside him. A man and a woman. The man was clutching a bottle of vodka, three-quarters empty. They looked rough, the sort of people you'd expect in a mugshot, captioned 'Wanted. In relation to kitten-torturing ring'.

The owner motioned for me to vacate the bed and move to the couch. Like it or lump it, or really fucking lump it, I was going to have roomies. The pair immediately collapsed onto the bed in a kind of urgent tangle, so I turned off the lights to show how seriously I planned to sleep. For the next hour they smoked and hit the vodka, each gulp winning a cheer and a burp and a cuddle. They tried to whisper, but they were far too sloshed to be tactful, and it came out as husky shouting. Eventually: snoring. And not your usual snoring: this was vodka-snoring, which is loud and much, much faster than you'd think it would be possible to breathe and not be dying of tuberculosis. I prepared for the final straw, and then I heard it: liquid splashing onto the floor.

'No-no-no-no-no!' I began, before recalling my place on the planet and translating to 'nyet-nyet-nyet-nyet-nyet!' I turned on my head torch, not knowing whether I'd be illuminating a man with his penis in his hand, or a woman, squatting. It was a man with his penis in his hand, of course. He hadn't even stood up, just turned to his side and fountained piss off the bed like a toppled statue. He then returned to sleep. In the light of my head torch, I appraised the flood damage. Thankfully, my stuff had been spared and he'd done a gratifyingly thorough job of pissing on his own shoes, which was probably a good metaphor for his life up to that point.

*

Kyrgyzstan began with meadows of wild flowers stretching to the horizon, where buttercup-yellow faced the blueness of sky. Herders cantered towards me and we passed the time of day grinning stupidly at one another. The capital, Bishkek, was supernova-hot and the air was still. Kids were playing in the fountains off Chuy street, running through errant sprays of cool water, and the national flag in Ala-Too square could have been a red dress on a hook.

I'd heard of a couple, Angie and Nathan, with a house in Bishkek. After opening up their garden for bikers, hundreds had passed through, camping on their lawn, pottering in the attached workshop, binge-eating, reading in hammocks and sharing stories, of course. If you've suffered on a bicycle, it's important to be vocal about this in front of other cyclists. The most transcendent suffering involves disease and natural disaster, and almost everyone had some experience of both.

The Kyrgyz postal system had coughed up a new Rohloff hub (a little metal home for my gears), after the last one had failed, and my next task was to build a new wheel for it. It was always easier to hand bike-fixing duties to someone more competent a mechanic than me, and ideally someone without such a chequered history of injuring themselves with multi-tools. Nathan was my man. I asked him if we could build the wheel together, and then I watched him do it while bringing him beers to encourage him. Nathan was a practical sort who spent a great deal of his time nailing something, climbing something or helping someone fix their bike, and when not doing that he was pitching questions about Hume's philosophies or trying to instigate a debate on the state of democracy. He was appalled at my ignorance when it came to bikes. 'This,' he'd say, pointing at my bike 'is not a sofa, you know. You need to—' I stopped him with a finger to his lips, handed him another beer and motioned for him to return to work.

My second major task in Bishkek was to sort out visas for the countries ahead.

'Ahhh, you from England!' said the man behind the counter of the snug printing and photo shop where I'd hoped to sort the visa preliminaries. 'You're like zero zero seven!'

I had to think for a minute.

'Double-O-seven? James Bond?'

'Yes! But maybe no. Your bike is mess. It look like some-sing from that film … Crrrazy Max.'

I took another moment.

'Mad Max?' I tried.

'Yes, yes,' he said impatiently. 'Mad Max, Crazy Max, it's the same thing.'

I'd once christened my bike Belinda, but the name was too homely for what my machine had become. 'Crazy Max' did more for the rust, cable ties and dents, so I went along with the transition.

Visas granted, I moved on again, south-west and upwards, to a grassland where herders peddled honey, fermented mare's milk and balls of strong cheese. The honey was magic and I poured it down like gloopy champagne. The following morning, I unzipped my tent on a scene of lush beauty. Beside the road, land fell away in a series of grassy natural platforms, vivid with yellow and purple wild flowers, and boulders of pink granite. Small coniferous trees of the sort you find in manicured country gardens were dotted about, their peppy scent in the air. Beyond the lowest platform was a streak of white water and then the land rose up again, flashing silver with wandering streams, and again, into mountains: massive, green and tiger-striped with snow.

Soon, the road was twisting beside the Naryn River, an uncanny green-blue. On the banks, family clusters trickled into one another and kids splashed about in the water or ripped into kebabs. The braver ones asked me a run of questions designed to bump up their sense of national pride and I was happy to oblige them.

Did I love Kyrgyzstan? ('Oh! It's incredible.')

Are Kyrgyz women beautiful? ('They really, really are.')

'Whatisyourname?'

'Stephen.'

'Like Steven Seagal!'

I was devouring a trout when an old man doddered over and asked for my name too.

'Stephen,' I said. He frowned. I decided to help him out: 'Like Steven Seagal.'

Unfortunately, he hadn't heard of the star of straight-to-video classics *Kill Switch*, *Flight of Fury* and *Shadow Man*. Also, he now thought my name was actually Steven Seagal.

'This is Steven Seagal,' he said to a young couple and their teenage son. It was too late to correct him. They smiled.

'Please, call me Steve.'

*

That evening, I sensed someone watching. I turned to my left to find a young boy in an Islamic prayer hat cycling beside me, openly gawking. For a while we rode idly together in silence.

His opener was a cut to the chase. 'Come and stay in my house!'

I thought I should probably get his name first. It was Ali. I was tired and Ali seemed nice, if a little over-eager, so I took him up on the invite. Ali lived with his extended family in a farmhouse covered in rugs. There were rugs on the floors, rugs on the walls, large stacks of rugs in case of a rug-less emergency or a sudden rug-failure.

'You have a big family, Ali!' I chanced, suddenly wondering if the sentence worked as I'd intended it to in Russian or if I'd just castigated his relatives for being obese.

Perhaps I'd be too if I lived on *plov* – an oily, salty mix of rice and lamb, the best comfort food on earth – which was rolled out and demolished. After dinner, I was swiftly dressed up in a traditional Kyrgyz robe and slept that night in Ali's garden on a rusty bed surrounded by the family's roving livestock, getting a late-night nuzzle, probably from a goat, though I couldn't rule out Ali's palsied grandmother.

I rode with Ali the next day too, and soon he was encouraging me to say in Arabic that there was 'no God but Allah and Mohammed is his Prophet'.

'Ali, are you trying to trick me into becoming Muslim?'

'These are good words!' he insisted.

'Ali, I'm an atheist.'

'Okay. But you believe in God, right?'

Ali peeled off before Osh. In the bazaar, I bartered for a few spare bicycle parts and food. With vast mountains ahead, grams mattered more than ever and I spent several minutes gauging the weight of various onions, eventually deciding upon the runt of the onion litter, a shrivelled excuse for a vegetable, which might actually have been a cancerous garlic. The man who sold me screws for my saddle was wearing a T-shirt that said 'property of the girlfriend'. It transpired that he was a strict Muslim with a wife and three children, no English, certainly no girlfriend, and little curiosity about what the slogan meant. I recalled other T-shirts around the world – an Ethiopian bus driver wearing 'I hope you like animals, because I'm a BEAST', and the stunning slogan on an oversized T-shirt worn by a girl in Yangon, whom I'm guessing had no English either. 'A blow job is better than no job.' A present innocently gifted by Grandma, perhaps.

The road climbed again beyond Osh, the days stiflingly hot, despite the altitude. I drained my two-litre water bottle every hour and searched in vain for taps or pumps for refills.

'Very clean, you can drink! Drink! Kyrgyz water very good!'

The old herder pointed at the river so I filled my bottle. But this was His River and I should have been more careful. Moving upstream, I found two donkeys pissing and shitting into the boil, and if you can picture a donkey looking enthusiastic, they were almost there. A little further up, a donkey had died in a watery purgatory, half in the river, half on the bank. My 'natural' mineral water was septic, donkey-brain soup. There's a moral here: never trust a patriot.

That afternoon, two kids dashed out from a yurt and gave chase for half a mile, staying on my tail like fighter jets, shouting: 'Bye bye! Bye bye!', delighted with the punchy iteration of the English words, even if they mistook their meaning. More boys grouped by the roadside offered high fives, which is apparently more fun if you do so with enough force to almost knock the strange man from his bicycle. But then you need some consistencies as you travel, some universal detail to keep you anchored and the world in equilibrium. Right then, it was the bubbly sadism of young boys. I recalled kids giving me the same forceful high fives in Turkey, Kenya and Peru. I felt nostalgic for the misfits with snowballs in Dartford. I thought back to the stone-lobbing kids of Ethiopia and that fat git who'd stoned me in Jordan while shouting 'You crazy donkey!'

The road dropped next into Sary Tash, whose residents were probably used to a certain dreamy distraction among their guests. The town was little more than a petrol station and a gathering of simple homes around a scissor of road, but behind it reached a grassy plain and beyond that, a wide, creamy band of mountains. It was my first glimpse of the Pamirs and it restrained me for a minute in the road. I longed to get stuck in, get climbing, ponder Sary Tash from a thousand feet further into the sky, no matter what that cost me in time and graft and hand-numbing high fives.

After my passport was checked at the Kyrgyz immigration post, I was waved into a rising no man's land where water the colour of clay jostled through a network of streams. And then a rock stood up. This was an interesting development. I stopped to watch it, wondering if the altitude was getting to me and if I should have a lie-down before the other rocks all stood up, donned sombreros and began some sort of cabaret. But the rock was behaving in a more inert fashion now, until it twitched, sniffed the air, scratched itself with a russet paw and disappeared into a hole. I was tuned in now – the valley flickered with marmots.

A marmot looks a bit like a squirrel that's eaten three of its friends. The wind carried their whistling calls to me. Over winter, marmots hunker down and chill, literally, letting their body temperature slip to almost that of the air around them, as their little hearts tick-tock along, five beats a minute. When the climate of the central Asian steppe shifted in the fourteenth century, and the ecosystem was disrupted, a bacterium that lived in the marmots and gerbils, *Yersinia pestis*, was transmitted to rats, then fleas, and after thousands of miles, territories and borderlands – humans. In terms of proportion killed, the

Black Death was the worst epidemiological event in recorded human history. We changed the world: that's what I saw in those pitchy marmot eyes.

14

Membranes

There wasn't a great deal going on at the Kyzylart Pass. A stack of empty switchbacks, a swirling, mineral wind. Hard to believe that this strange, spine tingling place was, in fact, an international border crossing. My passport was stamped in an oil tank, repurposed as an immigration office, and I was released into one of the oddest-shaped countries in the world.

Tajikistan has a wandering border of jags, tongues, legs and bite marks, more accidental splat than definite shape, like a dropped egg. By most accounts, Stalin, cackling like a cartoon villain, rushed off the national borders of the -stans with a pencil on a map, ensuring pockets of ethnic minorities in each territory and thereby discouraging unity and quelling old Soviet fears of an Islamic uprising. If not an outright myth, this is probably a prejudicial, Eurocentric reading of history, or at least an overly simplistic take on things. Central Asian borders were drawn with regard to census and economic and transport data as well as ethnicity, and had at times been altered at a local level rather than dictated by Moscow. We don't know, of course, how things might have been had the lines been sketched differently, but tension, like the sort that bubbles up every now and then in Central Asia (though arguably less than elsewhere with such diversity), is predictable wherever there is a line. Drawing borders is indelicate – like a surgical dissection – leaving scar tissue for life. There can be no authentic border just as there is no perfect wound: the blade won't slip in painlessly, following natural planes of tissue, it must sever and rupture and bleed.

The road beyond the border, scoring the Pamir plateau, was loyal to its online reputation. 'A little bit of everything' ensued: cracked and potholed asphalt, pools of sand, a cobble of rocks and cheap tar that melted and left oily tendrils on the soles of my shoes. Beside me were ridges coloured like blood stains, and I wondered if I was dizzy not for lack of oxygen but for the wonderful sparseness here, the sky of fast-moving clouds, their shadows marbling the plateau. I sensed a capricious land, history gashed and smashed into the mountainsides; the battle scars of wild weather, landslides, avalanches and earthquakes, a bygone violence that I felt as potential energy.

The M41 sounded like it should be a busy artery, but in fact traffic was scant. The road, better known as the Pamir Highway, was formally completed by Soviet engineers in the 1930s, though ancient armies had stamped its course for centuries, as well as Buddhist pilgrims, explorers and perhaps Marco Polo too, if the stories are true. Today, travelling cyclists come to the Pamirs each summer and, from a distance, they looked like pack mules. Closer up, the saddle bags turned to panniers, the hooves to wheels. Three lumbered towards me now, but they retained the mournful air of mules. Two men and a woman, dirty, wretched and grey-faced. They gaped at me and their eyes dribbled. There was a round of soft handshakes. I dropped my head, pressed on, as if towards a frontline.

I found Lake Karakul that afternoon, wind-crinkled and a lovely blue-grey. The mountains beyond, dull silver blades of rock, sliced into the sky. Before I climbed towards another pass, a woman stepped into the road. 'Here! I have tea. And apricots! Come! You're a guest! My house is your house! My food is yummy.' Giggling at her 'yummy', I followed her inside.

A television in the corner was showing what appeared to be an American medical drama. In a bleak, wind-smashed extremity of the world, the woman's teenage daughter avidly pondered the infidelity of a Texan surgeon, in subtitled Russian.

After Murghab, I turned off the Pamir Highway, intrigued by Lake Zorkul, source of the ancient Oxus, a river today known as the Pyanj and later in its course, the Amu Darya. I passed through Shaymak first, which flinched at the end of a valley below a mountain, and where life was clearly hardscrabble. There was a shop, almost bare, and a single donkey in the street, quite still, as if making a point.

After the village, the ground became spiked with brittle grass, and the earth, salt-stained, stretched away to mountains where it blended into the white of perennial snow. No vehicles passed me now and I began to worry that there would be a snap and buckle soon, some unfixable fault with my bike. I'd have to haul my stuff out of the mountains, eating moths and the straps of my sandals to survive. The road was stony and climbed steeply to several false summits. I got off to push, dropped my head, and trudged until my feet found even ground. Looking up, there was a large depression, half-filled by a lake and veined in silver streams. Around the lake shore, there were horns and skulls of Argali sheep, an endangered mega-sheep, weighing up to 350 kg. You can blast one into eternity for $30,000 if you're so inclined, and plenty of moneyed American males are.

I rounded the lake and found a trail through a baize-green valley, no herders here, but plenty of that delectable out-of-my-depth feeling. It was rapture, being alone in some small, spare part of an unknowable place. I shivered with a kind of welcome dread. Edward Abbey nailed it in *Desert Solitaire*: 'I am twenty miles or more from the nearest fellow human, but instead of loneliness I feel loveliness. Loveliness and a quiet exultation.'

A few yurts marked the entrance to Zorkul National Park where a family were lassoing yaks in a pen, cutting the wool of the adults and tagging the young, a frenzied sport of horn and leg grabbing. The father of the family guided me to a yurt insulated with wool and brought me meat soup, curds and bread. I'd zipped myself snugly into my sleeping bag when the door opened and his teenage daughter raced in. She was holding a phone. Placing it under my nose, she whispered 'This is I!' On the screen was a selfie, not of the girl beside me in a headscarf and old-style Kyrgyz dress, but the same girl, eyes narrowed, head tilted, in lipstick and jeans. She snatched it away and rushed out of the yurt. There was something so moving, so hungry about this, that I was kept awake that night, lost in thoughts of the contagious quality of culture, of how ways of life will always diffuse and mutate, building and wrecking and remoulding the world as they go.

I lost the trail the next day and, beneath a huge cloud of black flies, followed a line of broken telegraph poles, eventually finding the Pyanj River as it left the lake. Towards the close of the nineteenth century, amid the colonial scuffling known today as 'the Great Game', a frontier was drawn along the Pyanj River, leaving a strip of Afghanistan – the Wakhan corridor – as a buffer to the British and Russian empires. As I cycled beside the Pyanj now, I wondered how many international borders were as patently divisive as this one. My side, the Tajik side, was scattered with homestays, grocery shops, even the odd ice-cream parlour. Afghanistan, from what I could discern, was a few donkey lanes and mud shacks, yet on both sides of the river, and that invisible man-made imposition, people spoke the same language, practised the same religion and shared the same ancestors. Predictably, official divisions had fostered others, and the relationship between Afghanistan and Tajikistan was tense at times, the former weathered by

Islamic extremism and more than thirty years of conflict, the latter anxious about refugees, militants and drug smugglers sneaking across the river.

At a web of thrashing streams gushing into the Pyanj, the Soviet-era bridges were warped scrap metal, making rapids of the water. I heaved my bike over the streams and then, from somewhere, a voice rang out. Looking up I saw a military watch-tower: the way ahead, I noticed, was gated. This was the military base at Khargush. Two soldiers made their way towards me.

'You! Where you come from?'

'From Zorkul,' I said.

The captain arrowed in on me. He looked more Russian than Tajik, wore wrap-around shades, a cap and army fatigues. A finger remonstrated with my face.

'Border area. Terrorists,' he said, slow and bullying. 'You've been to Afghanistan.'

'No no! Just Tajikistan.'

'Documents!'

I gave him my passport and the permit I'd bought from the father of the family near the lake, which granted me passage via Zorkul. He seized it immediately.

'Counterfeit!'

I'd guessed that much, but the family had doted on me, and the fake permit only cost five dollars, which seemed a reasonable way to repay their hospitality. 'Search him,' he said, nodding to a soldier.

I'd been robbed elsewhere by police on the pretence of a search so I declined the kind man's offer before quickly realising that this might not be a great time to discuss non-searching options. 'You're waiting for my ID?' said the captain. 'Here's my fucking ID,' he pulled a handgun from his belt and for a moment I thought he'd aim it at me, but he lost courage, returned it to the holster, and then made a gun with his index finger and thumb, put it to my temple, pulled a fictive trigger, and my brain was splattered into the dust.

Soldiers kindly unpacked my gear, flinging it piece by piece into the trail, the captain idly kicking the back wheel of my bike, over and over, whistling as he did so and glaring into my soul. Apparently, my soul gets boring after ten minutes or so, and he loped off with a sneer for goodbye. I was now free to leave, this message conveyed by a young soldier who returned my passport with an armful of freshly baked bread. I was beginning to see a pattern in my treatment by authority figures throughout the world, as if they all had the same secret checklist: detain, interrogate, search, threaten with deadly force, feed, bid cheerful goodbye. Tick, tick, tick.

I was juddering along a corrugated road now, a relief, since my standards were rock bottom after dragging my bike across a roadless, fly-infested wilderness. Perhaps this was why something extravagantly bad was happening to my arse. I couldn't put off an investigation any longer, and without a mirror, or an extremely close friend, there was only one thing to do. That evening, in my tent, I reached for my camera and held it behind me for a close-up. I looked at the viewfinder, tilted my head one way, then the other, turned the screen upside down. This was not normal anatomy, and it's incredibly hard to describe what I saw, but I'll try.

There is a species of sea anemone called *Actinia equina*, the beadlet anemone, common to the shores of the British Isles. Know it? A liver-red, glistening thing that wobbles like blancmange. Now imagine a pair of them, only greatly increased in size, and there you have my bum. I was therefore immensely grateful when the road through the Wakhan improved in sections the next day, allowing my arse to begin a slow convalescence, at least I supposed so. I never had the heart, or indeed the stomach, to photograph it again.

There was just a trimming of green left to the river now, and a few weedy camels had the far bank, where Afghanistan, steep and dry, drew itself up into snowy steeples of rock. I stalled in the road, stunned by the Hindu Kush, a 500-mile-long body of mountains that fans deeper into Afghanistan. Its history is bloodied by military conflict and the ancient trade in slaves, seized in South Asia and marched over the mountains to the slave markets of Central Asia. Another stain on the story of the world.

Streams crashed through Langar, the first village I'd seen in five days, trees tracking them, making verdant laces of the land when viewed from above. I descended through a snow of poplar fluff, then continued beside the river to Khorog. I'd been invited to visit the Aga Khan Foundation, a popular force for good in the region, and one that invested in education and bridge-building, of both the literal and figurative kind. Five roads now traversed the Pyanj near here, connecting Tajikistan and Afghanistan, and a cross-border healthcare project had been set up to aid Afghanis on the rougher side of the divide where there were rural doctors but no medical specialists, and where communities grew especially isolated come the winter snows. Under this system, Afghanis requiring urgent or specialist treatment could sometimes get it in Tajikistan without a passport, a pragmatic, transnational response to difficult circumstances. The project was trialling e-health too, whereby patients were assessed by video call when there were no experts locally and the Afghan rural doctors could get advice. Complications involved a paranoid Tajik government, sometimes shutting the programme down, anxious that more than health advice was being dispersed, and language difficulties: Tajik doctors had generally trained in Russian, Afghan doctors in English.

Reports were coming in of an Afghan woman with a severe

head injury following a road accident, deeply unconscious. I set off in a car with the mousy-moustached Dr Umed who took referrals from Afghan doctors each day. We were not planning to cross into Afghanistan ourselves, but from the border we were picking up a Tajik urologist, who'd been working in rural Afghanistan, and delivering an anaesthetist and a surgeon, the latter wielding a drill with special attachments in case he needed to make a hole in the woman's skull to reduce the pressure inside. This was street-fighting medicine.

At the checkpoint, Dr Umed returned to the car, looking downbeat.

'Soldiers say the situation bad – they're closing the border. There is a new militant group shooting in Iskashim. Tajiks are being evacuated. We must wait.'

'How serious is it?'

'Not sure. Some Afghan government troops are defecting. You know Afghanis can all use AK-47? From little boy to old man. And police, how you say, don't mess about? They shoot to kill.' He sighed. 'And they shoot a lot.'

But a minute or two later, the soldier returned and appeared to relent.

'It's okay,' said Dr Umed, opening the back door. 'We have permission. But we must get the doctors back today, they won't let them stay overnight.'

Our two doctors walked across the border, and the urologist climbed into the back of our car. 'This man is a great urologist,' Dr Umed piped up. 'The best in all of Tajikistan. And not only that, but he is a good, no, no, a *great* man too. One of the best. A great, great man.'

This was an agreeable Tajik habit, I'd begun to realise: gushing compliments for friends in public. Outside Khorog General Hospital, Dr Umed exchanged 'salam ali's with twenty people, touching his head to theirs, holding his hand over his heart, bursting into, 'This is Dr Mahbut, he is a great man! An incredible surgeon, the best we have. A genius!', and then, 'This is Nizoramo, she is wonderful and will look after you. She was my colleague in medical school. Very wise, very wise. And brilliant! A lovely woman! She knows everything, she will help you.'

Khorog hospital had the blended aromas of every hospital, poo and disinfectant winning out. A cat scampered past the bed of a nineteen-year-old boy who'd been shot in the abdomen in Afghanistan. A surgeon in blue scrubs laid a hand on the boy's shoulder.

'We don't know why,' he said. 'The Afghans ... they don't always tell us. Maybe family feud. Maybe local politics. People get shot in the bazaar very often. Perhaps he was even shooting at the time. We don't know and he won't say. No point asking.'

We moved on, but not before the boy had kissed the surgeon's hand.

*

A few days before I'd arrived at Khorog, violent gales had wracked the Pamirs. Fifteen kilometres east of the town, an enormous hunk of ice high in the mountains had begun to melt in the special heat of that summer. Then gales freed it. A huge mudslide dammed the Gunt River and forged a rapidly swelling lake. Around 80 homes in the village of Barsem were washed away and many more were evacuated in the risk zone. A power station serving thousands risked submergence. Quite suddenly, Tajikistan was the place to be for international crisis experts and they were flown in from all over the world to monitor the situation and advise. We drove up to a camp for people made homeless by the landslide, on a platform of mountain, previously a military site, where the ground was still thick with old bullet casings. Tents from the Red Crescent stretched out in rows, and evacuees milled around.

No one had died in the flooding and by most accounts the emergency response had been swift and coordinated. There was a lethal tragedy unfolding below us though, and it was being given far less attention. In Khorog, opium frequently claimed the lives of addicts, but the death toll was rising steeply of late. For years, the border between Afghanistan and Tajikistan had been firmly controlled by the Russians; Afghanistan was less penetrable then and there were relatively few addicts in Khorog. But in the wake of independence and Tajikistan's civil war, the border control went back to the Tajiks. The no-nonsense approach of the Russians was history. Now that the border was controlled on both sides by people who spoke the same language, corruption was a much slicker process. Opium began to flow more easily and the road to addiction is quicker with purer gear. For the hundreds of addicts in Khorog, the funding for treatment and recovery was pitiable. I met some of them at a treatment centre, and while most were reticent about talking to me, the stories I did hear grew familiar. The men had been using in the military. They were divorced, had hepatitis, worked occasional short contracts in construction. Their brother was an addict, friends had died or languished in prison. Meanwhile, the press and politicians painted drug use, like migration, as a plague, not a tragedy, and likened addicts themselves to germs.

Maybe borders are less like scars than I'd thought, and more akin to cell membranes. A scar implies permanence and gives no sense of the fluidity of national lines in the long term, or the fluctuations in their power. Cell membranes, on the other hand, are semi-permeable, allowing certain solutes, molecules and substances to pass freely into cells, blocking others. Borders have this trait in common too, permeability varies, depends on what's crossing, and plenty tries: opium, goods, people, aid, multinationals, medicine, infections, money, knowledge and ideologies. Of course, we wouldn't exist without cell membranes, and I wondered now if it was right to blame borders for all the frictions of the world. Perhaps our reverence for them, or our hesitance to bridge them, mattered more.

*

I left Khorog with James, a young and upbeat British cyclist whom I'd met only the night before, in a campsite. We were only a day out of town when a different breed of soldier appeared on the road. The Tajik border patrols I'd seen so far were boyish men, trotting along in untidy formation, drowning in outsized camouflage gear. This troop waded along the road, all scowls and biceps, squeezed into the jet-black get-up of special ops. We saw more soldiers that evening, lolling in mosquito nets, giant machine guns trained on Afghanistan. We were nearing the Taliban strongholds of Kunduz and Faizabad now. Three years later, a group of foreign cyclists were attacked by militants here: four were murdered in the road, run down, hacked and stabbed.

We were squeezed now, the river on one side, a steep mountainside scattered with boulders on the other. Camping with discretion would be tough, but as dusk edged in I spotted a steep trail and made a quick recce. It led to an abandoned hideout for the military to watch Afghanistan, and a car-sized boulder provided some shelter. We settled in.

This was the art of 'stealth camping', or at least it should have been. Our main problem was James's tent, which was an intense yellow and had all the subtlety of a flair gun fired into an oil well. Nothing could be done of course, so we ate hunched behind the stone wall, and then I zipped myself into my tent.

About a foot of zip on the inner tent was broken towards the top, but I figured I could leave it open; it was too dry for mosquitoes. As I lay back, postprandial and burping, something caught my eye. A shadow, the size of a fist. It whipped across the net inner, leapt through the gap in the zip and landed – pmff – on my thigh.

I now know the sound I'll make in my final moments if I ever meet a violent and untimely death. It's a theatrical, quivering trill, in the vocal range of a glam-rock front man. I snatched for the nearest weapon – a sandal – and began slapping it uselessly around my tent, whacking my own legs with the violence and tempo of a drum fill, screaming in fear at first, then self-inflicted pain. The thing dashed – I swung – splat. Crawling over to it, I peered down at the most appalling aesthetics of a spider and a scorpion rolled into one. It was big, desert-coloured, with unmissable fangs. I could hear James shouting 'What is it! What's happening!' He poked his head inside my tent and we both frowned over the remains of the camel spider. A misnomer really, for they are not technically spiders, nor do they murder camels, as you might reasonably assume. There are all sorts of other myths around them: they do not scream like a banshee, for example, but they do have a fetish for leaping into shadowy spaces (the order is solifugae, Latin for 'those who flee from the sun'), which means they are often perceived to chase people (they're just trying to find some shade). Even so, I slept with a sandal in my hand.

The next day we continued downriver; the waves and eddies gave off a cool radiance. The sky was crammed between rock, an incidental strip of blue. Up and over the Khaburabot Pass, where signs warned of landmines, we began following a new river, the Obikhingou, as grey and trundling as cement. That night, camped on the banks, I was reading in my sleeping bag when the night was ripped apart again.

'Fucking giant scorpion death spider!'

I looked out. James was giving a credible demonstration of the *hopak*, an ancient dance, still practised in Ukraine, which involves some frenetic jerking of the legs while in the crouched position. The illusion was broken when he began hitting himself in the ears and screaming, 'Please, oh God, please!' A monstrous camel spider was scuttling around the sand at his feet. I wiggled out, prowled a bit, took aim and slapped it to goo with my sandal.

James was red-faced, woozy and, it seemed to me, about to pass out from hyperventilating. It was his turn to cook, so he zipped up his tent until just a small section of the door was open at the top, from where a pair of wide eyes peeped out. With his arms hanging limply through the gap to his elbows, he tried to chop an onion, which he succeeded in doing, but only if you consider a chopped onion to be two bits of a whole onion.

We came across a Polish cyclist on the road the next day, heading in the opposite direction. He looked troubled, and perhaps life would have been easier if he wasn't so insanely loaded. He had four stuffed, deep panniers, several dry bags and an overloaded trailer. His repair kit would have been useful if he ever stumbled upon a damaged aircraft carrier, and his tent could have been blagged from the Red Cross. Tentatively, I asked him how he'd been enjoying the cycling so far.

'I fucking hate the children. All the time saying "hello hello".'

'You hate the children saying hello?'

'Fucking hate them!' He raged. 'I say hello once already! Why you ask me again and again and again! I need hotel. This is awful. In my country, you camp in fields like this, someone rape you with pitchfork.'

'It's fine,' I assured him. 'I haven't seen any pitchforks. Saw a few guys holding staffs.'

'Yeah,' said James, 'that wouldn't be as bad.'

The Pole glared at us. 'And this road terrible. How will I cope with this! My ass and balls are really suffering, I tell you.'

I looked up and noticed that this track had nothing on the ruts and potholes he had to come. I had a sudden urge to lay a hand on his shoulder, look him in the eyes and say, 'It gets worse, son. This is a fucking holiday. You're gonna wish the road was this good in a few days' time when you're up to your balls in camel spiders, can't walk without wincing and have to photograph your own arse to find out why.'

I just said, 'Yeah, it's a bit bumpy.'

It was an ugly few hours on the road to Dushanbe. In an expansive mine, machines threw pale dust into the sky. The traffic picked up in grunting waves. As the white Land Cruisers of NGOs zoomed past too close, I realised that if I was to meet my end in Tajikistan it probably wouldn't be at the fangs of a camel spider, or the rifle of a soldier, but under the wheels of a humanitarian.

In the dust, James disappeared somewhere behind me. He reappeared ten minutes later wearing a kind of plastic visor.

'What's that?' I asked.

'Sunglasses. I lost mine, so I made these. I can't see anything in this dust.'

'How did you …?'

'Fanta bottle.' I could see that now. He'd knifed up a Fanta bottle and attached it to his face, looking very much like he was off to a Star Trek convention, but in an outfit fashioned by his elderly, cohabiting mother. I was going to miss James.

The president, Emomali Rahmon, welcomed us into Dushanbe from mawkish posters on billboards. He shook hands with the working class, hugged religious leaders, held grain, and, my favourite, waded through a tide of poppies in a suit. Here James peeled off, and I headed west towards the border with Uzbekistan.

Rumour had it that long, drawn-out searches complicated this particular border crossing. Time spent searching was an investment: it produced more contraband, and contraband meant good money. Even finding prescription painkillers like codeine would mean a handsome payday for one lucky official, and, if not, twenty minutes in the joyless confines of an Uzbek jail, staring numbly at its bucket-toilet, would convince even the hardiest traveller to flop out the contents of their bank account. The officials were generally young men who focused particularly on searching hard drives. I assumed that there were three main reasons for this: pornography is illegal, so finding some would mean a windfall; pornography is otherwise hard to come by for young, horny, unmarried Uzbek immigration officials; and they were bored.

There were no vehicles waiting at the border and I had a doomed sense of what was to come as an official held up a hand. I handed over my passport, and then hung my head and waited to be asked to undress slowly, sing like an Uzbek woman and dance for them.

'If I want a ticket for this Manchester United, how much it cost?'

I made up a number that seemed unreasonable and probably true. He made the appropriate gasp but didn't yet let me proceed. Instead, he stayed inspecting the visas in my passport.

'This one?'

'That's for India.'

He nodded.

'How is the Maharaja?'

'He's fine.'

There were some notable low points over the next two hours. The methodical palpation of the lining of my head bag. The ten minutes he spent peering into every individual section of every tent pole. He broke open my bread rolls with his fingers and peered suspiciously at the crumb while three others bunched around a monitor and began searching the contents of my hard drive. I imagined that they would soon stumble across the photo I'd taken of my enflamed arse, instead they discovered the film adaption of Orwell's *1984* and fast-forwarded to a scene in which a nude Julia (played by Suzanna Hamilton) throws her arms open to embrace Winston (John Hurt). They paused the naked actress and tutted, almost as if they meant it. I sat, fretting. I'd never live down the irony if I got thrown into an Uzbek jail for possessing a film about totalitarianism.

But they let it slide and I was a free man, released into the dark flatlands of Uzbekistan to the ring of cicadas. A half-moon lent me a slight silhouette, and shouts of 'Otkuda?!' (Where are you from?) were pitched from the dim shapes of people near the road. I shouted 'Anglia!', and the word was swallowed for a moment by the night then returned with 'welcome!'

For two and half days I watched cotton plantations turn to cabbage fields to sugar cane and back again, spending my Uzbek millions on melon and round bread ($50 equalled a 5 cm-high stack of Uzbek som and supermarkets came with note-counting machines). The days were clear-skied now and a dry wind was blowing as I reached the southern city of Termez, gateway to Afghanistan.

15

Dislocated

Afghanistan made for grim reading on the travel advice section of the UK Foreign Office website. 'Multiple threats', 'extremely volatile' and 'consider using armoured vehicles' were just a few of the phrases that piqued my terror. The whole country was painted a deep red, the colour of oxyhaemoglobin, or of fresh bullet wounds. So in Termez, I'd arranged to meet Sam, the Australian cyclist I'd stayed with in Xinjiang. It would be nice to have a companion, and in the event of a rampaging assault by Taliban fighters, he could be a serviceable human shield too. I decided not to mention this.

This was a short departure from my westward journey home. My plan was to ride south to the desert city of Mazar-e-Sharif, a journey of only one hundred kilometres, double back to Uzbekistan and then start a long ride, north-west, to the Caspian Sea. Pedalling beyond Mazar would be unwise, to put it mildly. It had been a bad year in a run of bad ones for Afghanistan and the north of the country in particular. The Taliban controlled around a third of the country now, more than at any time since the American invasion of 2001. Mazar was considered safer than most Afghan cities but there had been targeted massacres here recently too. A few weeks ago, nine workers from a Czech NGO, People in Need, were murdered in their beds, and a few weeks before that, the Taliban had stormed the provisional prosecutor's office in the city centre at lunch time, an escalation in force and audacity that few had anticipated, portending, perhaps, a new dawn of Taliban attacks. Few people knew the true number of dead: the Taliban gave high numbers, while the forces defending the city, loyal to the provincial governor, played down the death toll.

Sam had no intention of being voted an infidel and popped off from long range. He'd gone full Afghan: a beige shalwar kameez, a *shemagh* around his neck and a pair of leather sandals. He'd even grown a bushy beard and shaved his moustache in the Islamic fashion. Taking in this new Sam, I couldn't help but feel that his efforts to blend in were offset somewhat by his loaded touring bicycle and luminous yellow panniers, his eyebrow and ear piercings, and the tattoo of an open skull on his calf, the logo of the band the Grateful Dead, which slipped free of his shalwar kameez as he pedalled. If he did get shot, saving me from the same fate, perhaps I'd draw comfort from the irony of the band's name. I decided not to mention this.

From Termez, the road south hopped the Amu Darya River with a single-track railway. We crossed the border and started out over the militarised, unfriendly-looking bridge: 'the friendship bridge'. Halfway over, Sam got a puncture. Crouching low on the concrete, glancing sideways at the sand beyond the river bank, we teamed up to make the repair, working fast.

I gestured at Sam's eyebrow piercing.

'The Taliban hate those. We get abducted, you'll be executed first.'

Sam snorted, sanding the inner tube more viciously now. We carried on like this as we reloaded Sam's bike and pedalled over the bridge, but behind the gallows humour was a nervousness about what lay ahead and, for both of us, a reluctance to face what had drawn us here. Books had deepened my curiosity for Afghanistan ever since I'd walked with Eric Newby through the Hindu Kush, but nosiness wasn't my only incentive. Over the last year, I'd been pulled insidiously towards more remote places, and I enjoyed feeling on edge and exposed. I liked the extremism, the sense of control drifting away. Like Turkana in Kenya, Chin State in Myanmar and western Mongolia, Afghanistan was one of those unlucky infatuations. It felt out of bounds. Perhaps this was the same thirst for trouble that wipes out free soloists, polar wanderers and those oddballs who keep tigers as pets, though this was something I considered more in hindsight than at the time. Staring into the sands of Afghanistan, I didn't question myself too much. Perhaps I was afraid I'd be dishonest, or outed as reckless, or that I didn't know enough to explain the choices I was making.

Unlike the desert of Uzbekistan, which was held at bay by irrigated, arable land, only sand prevailed here, and sometimes dunes imposed on the asphalt, forcing us into parabolas. Landmines exploded from time to time out there in the sands. I searched the horizon, spooked by the intensity of the sun, the raging space, the very notion of Afghanistan … so wild, mysterious, spiced with threat.

'Hey guys! Guys!'

A driver had pulled alongside and was leaning out of his window. I tried to ignore him, refocusing on my clumsy, overheated fantasy.

Just the name … Af-ghan-i-stan …

'I'm from Slough!'

'What?!'

'Slough! Many years my friend! Ha ha! So nice to see you in my country!'

I glanced past him, but there was no jihadist in the passenger seat, no ISIS bumper stickers, no bug-eyed warlords lurking in the footwell, and with a strange calm I wondered how and when the rocket propelled grenade attack would begin.

'You know this Slough?'

I looked at him again, and he looked at us, and a horn from an approaching truck reminded us that nobody was looking at the road. He veered across the lane, wheels roared past, and he veered back towards us again. It was clear that he wasn't going to go away and leave me to brawny adventuring.

'Slough! Like on the TV. Like *The Office* with Ricky Gervais! David Brent! David Brent!'*

There are many ways to exemplify the cultural and economic faces of globalisation: China's New Silk Road, a McDonald's in Vietnam, the explosion of Indian call centres. But this is my personal favourite: a man screaming David Brent out of a car window in a South Asian war zone.

Eventually, he sped away, leaving us to a much tamer desert. Still blisteringly hot though. I reached down and jerked my water bottle from its holder, tipped it to my lips and watched the last rivulets run down the inside of the plastic. Sam was leaning heavily over his handlebars, looking sickened by the heat.

The road arrived at a T-junction where lorries were parked up, their drivers unfolding mats or praying in boxes of shade. We were already decided on a right turn to Mazar-e-Sharif, but if we'd had a brain fart here and swung left, we'd be on our merry way to Kunduz, which skirted territory under Taliban control. (In one month's time, the Taliban would storm Kunduz itself, taking the city from three sides and setting Taliban fighters free from the jail.)

Mazar was a mere 20 kilometres away now, but the road was strung by police and military checkpoints; real ones, I hoped. At one, police moved in around us, muttering, unsure

---

* For my international friends: Slough is a depressing English town/post-industrial wasteland 20 miles west of London, famous for trading estates and for being David Brent's fictional home in *The Office*.

how much they needed to say about our gamble and how much we might already know.

'Taliban,' said one, glancing around with wide eyes to convey their imperceptible ubiquity. Another simulated a beard with a drop of his clawed hand below his chin, and a turban by turning circles with his finger above his head. 'But we'll protect you,' he explained, pretending to shoot into the desert, with a grin that inferred bloodlust.

The edge of Mazar was all smoke and littered car parts, small industry grinding away for a population on the rise. Violence had spread through nearby villages, displacing people who'd turned to the city now for protection. A little reassurance, I supposed, buzzed above – a US Chinook helicopter, a relic of a fading international mission, and to the south I could see an airship, which did for floating surveillance. This blatant militarism clashed constantly with the spectacle of men shouting welcomes to us and waving hysterically from cars.

At the heart of Mazar there was a square of road around the shrine of Hazrat Ali, a grand mosque tiled in aqua and diversities of blue. We stopped beside the first hotel we saw, and from the lime-green lobby I deduced a low bill and a shit-smeared squat toilet. We got large burn marks as well, which seemed to suggest exploding electrical appliances or impromptu arson. The owner looked as if he'd just received a heavy blow to the head; all day he sat by the window grinning, stupefied by the city, smashed on hashish.

We quickly figured out our room's idiosyncrasies, unscrewing the light bulb to turn it on and off and twining sparking wires to work the ceiling fan. Through the window, we had a pleasingly wide view of the mosque and the streets below. It was a three-stage jump, 10 feet or so to each balcony, then pavement, and I found myself suddenly appalled that I'd needed to plan such an inevitably crippling escape route.

The next morning, half asleep, I heard the sounds of Mazar flow in through the window. Men were calling, and then, a small fracas: I lifted my head to see a tough gang of street kids tussling over the fruits of begging. The emerging sun restored colour to the domes of the Blue Mosque and threw the courtyard into fresh white brilliance. A scattering of women wandered about on early errands, sky-blue burqas rippling in the desert wind.

As Sam slept, I watched vendors gather, greeting friends and getting stuck in those interminable handshakes of South Asia. I notched up usages for the *shemagh*, the Afghan scarf. Often, it was simply a shield from the sun, but it was also used to carry melons, to swat at flies, to sit on, to steal from Dad and whip your brother with, and, more than once, as dental floss. But of all the early morning comings and goings, it was the trucks that seized my attention. They dragged long shadows up and down the square of road that enclosed the mosque, with gangs of men sitting in the open-topped backs, slung with silvery worn assault rifles, legs hanging over the side and *shemaghs* wrapped around their heads and faces, leaving just a slit for the eyes. Wraith-like men, with mounted machine guns. Some were police, others paid militias, at least I hoped so. When the Taliban had attacked the city, they had done so in similar disguise.

The militias were loyal to Atta Muhammad Noor, the famously wealthy governor of all of Balkh province, principled strongman or violent warlord depending on your politics. Known as 'the teacher', Noor was a former commander in the Mujahidin and the job of governor, he'd boasted, was his by rights, unconditional on the mere opinion of the president of Afghanistan.

Sam and I started the day with a stroll, browsing stalls around the square where men sold everything from sandals made of car tyres to screwdrivers with stars and stripes handles. One began shouting 'Bicycle! Bicycle!' – to Sam's dismay, shalwar kameez or not, Mazar was well informed about who we were. Two carpet sellers were making a pedalling motion in the air with their hands in case not everybody had figured it out. I hoped, though it felt self-aggrandising, that foreign travellers like us could be seen as a good omen for Afghanistan; perhaps that explained the odd smile. I was a voyeur as well of course, looking in on somewhere part worn by violence: I couldn't be here and avoid that. But most often I felt like a guest. Hundreds of people welcomed me cheerfully that morning. Only twice did I catch a whiff of something else, something not friendly, the men undisguised about it, though they remained aloof and might simply have burnt their breakfast or slept through their alarm, you'd never really know.

We sat below the hotel in a cosy eating house where men worked out front, their thick, practised forearms twisting and digging into ice cream. Women ate in a curtained-off place at the rear and above us a muted TV set threw out images from battlegrounds: tanks hammered through the desert, military commanders addressed the camera, soldiers ducked, ran, signalled, ducked again. Shots fired, shots fired, shots fired.

But there was little chance to catch the action as Sam and I were soon surrounded by fluffy-moustached students. 'Are you American soldiers?' asked a boy who introduced himself as Hashmat.

'Just travelling.'

He nodded.

'We are tired of war,' he said, flicking a hand at the TV as if to dismiss it from my attention. He whipped out a phone and asked for my Facebook page. I spelt it out for him and watched as he went on a liking frenzy of my posts, working his way back to 2012. Hashmat still messages me from time to time, and I message him, trading snatches of our discrete lives.

Three young men sat down with us too, translators for the US military. Naser wore a T-shirt with an American eagle and the words 'US ARMY', which turned out not to be a perk of the job but a souvenir from the only foreign holiday he'd taken in his life, to Goa. But they were marooned by conflict now and rarely left Mazar, even for nearby Balkh.

Together, we left the ice-cream shop and wandered past the mosque. Nasar was twelve when the Taliban took the city, the spoils of a third attack and the beginning of three years in power.

'The Taliban would leave their weapons here when they went to pray.'

He pointed to a few steps beside the mosque.

'Nobody would touch them. They could just come to your house, any time, day or night, and you had to feed them.'

We walked on, past a school, its walls topped with barbed wire.

'This was their last stand. It has a basement, so the Taliban took shelter there. I remember helicopters from America in the sky. The next day, I saw body parts in these trees.'

*

I'd convinced myself that Mazar would be largely off the travel-networking grid but the next day Sam had arranged a lunch date after connecting with Wahed online. We found him sitting on a bench outside the Blue Mosque. He was a science professor at a college in Mazar, a soft-spoken Hazara with a formidable brow. He was related, in some convoluted way, to Abdul Ali Mazari, a Hazara politician who'd famously opposed the Soviets and campaigned for a Federal Afghanistan before the Taliban arrested him near Kabul in 1995. He was tortured before his execution.

'Today is national holiday for our so-called independence,' he said, a little antsy. 'Many people out today, you see, the city is like a museum. You'll find Hazara like me, Pashtun, Uzbek, Tajik. There are Turkmen. This is not like Kunduz. I've never been there, and it's not far, but they kill Hazara like me there now. I suggest only one thing, when you're here, don't tell people your plans. Especially the Pashtuns, they may have links to Taliban.'

A few women approached us, unveiled, colourfully clothed, with nose rings and studs, as conspicuous as flowers in the field of blue burqas.

'They are *Fallben*,' explained Wahed. 'Fortune tellers. Want to know your future? Come on! We'll ask.'

Frankly, within the turmoil of my own frightened existentialism, that sounded tempting. Wahed gave one woman some money and she took a quick look at our palms in turn, reciting thoughtlessly fast her pick of futures for us, with Wahed translating. Sam would travel soon and marry within one month. I was going to father eight children (five boys, three girls).

A car passed, beeping. 'My cousin,' said Wahed. As we walked, a man passing on a motorbike honked his horn too. I looked at Wahed. 'Another cousin,' he muttered.

'You have many cousins?' I asked.

'Oh, around 200 or so. My grandfather was a polygamist.'

There were more relatives to meet that day and we drove off to find an uncle. Abdul had thinning, grey hair but eyebrows as dark and bold as his nephew's. Like Wahed, Abdul had fled to Iran to escape the Taliban, returning after the Northern Alliance retook Mazar.

We sat in comfortable silence at first, propped against cushions, no pressure to entertain or compete – unlike conversation between travellers – picking at Bombay Mix, sliming our faces with slices of watermelon and sinking endless top-ups of black tea. Abdul's son brought in rice and chicken and I watched Sam get frantically stuck in. He was on the rebound from years of vegetarianism and had eaten little but meat for weeks.

Abdul smiled at Sam's indulgence. 'How is your father? Is he in good health?'

Sam looked taken aback. 'He is.'

'And in Australia, do you have melons like this? How big are they?'

Sam threw open his arms to invent a monstrous, world-record melon. As our laughter died down, Adbul brightened.

'Ah, I remember now! In Iran we watched this TV show from Australia, such a good one. The protagonist was a kangaroo and his name was Skippy. Do you know this Skippy?'

When Sam snorted 'yes!', Abdul said simply 'We have very different places, you and I. Two mountains cannot reach each other, but two men can.'

Abdul talked about a constrained life in Iran, his arms playing in the air, like he was hanging up his words on an invisible line.

'I don't have good memories. We were refugees, you see. We couldn't buy a car or own a house. And the lower classes from Turkey and the Kurds were very resentful towards us. Us Afghans ... we will work for less.'

Wahed jumped in.

'We shouldn't have to leave our country. The problem has always been Sunni Islam. The Sunnis ... they are ISIS. They are Taliban.' His face tightened. 'They murder for God.'

Disgust ran through him now, and in that snatch of hate I felt suddenly conscious that I only knew the history of Mazar from a few paragraphs. I hadn't lived it. There were forces shaping these animosities well beyond my ken. The Hazara were singled out when the Taliban attacked Mazar in 1998; there was torture and rape. Then more tit-for-tat violence, for which you could never really find an origin, each atrocity a retaliation for something else, each horror spilling into more. The conflict couldn't be as uncomplicated as some western commentators made it sound, with their talk of ancient hatreds between people – that sounded to me like an easy exposition, and an easy way to kill all hope of a resolution. And conveniently, it sidestepped the issue of foreign powers arming some Afghans over others, manipulating the hatreds and fuelling the fire.

We left Wahed and got online, the internet café a laptop balanced on a cardboard box on a street side. The Afghan ISP triggered a salvo of adverts sponsored by the Australian government explicitly telling Afghanis not to leave their war zone and make the journey to Australia. The Aussie government had sponsored films about refugees too, with people acting out years in detention and rejected asylum applications, financial ruin, death at sea. Nobody in these films had a backstory that might include a brother, cousin, friend murdered. It was hard to see how young Afghanis would associate Australia with something as benign as a jolly kangaroo these days.

*

Wahed had left me with the name of a local physician and the next day I looked him up. Dr Ali worked in the city's main hospital. Men stood by the doors, apart from the women, and there was an array of hats, skin tones and faces, emerald and café-au-lait and blackish eyes. Mazar had the multiculturalism that I'd assumed the badge of western cities. The notion that tolerance is a uniquely western virtue felt like a fairy tale here, despite the surrounding conflict.

Dr Ali was an orthopaedic surgeon, short and authoritative, who took firm hold of my hand in the hospital corridor.

'Come on, I'll show you around. You'll see we have plenty of problems here. The medical schools for a start. The doctors come out with little clinical experience; they are trained in one specialty by specialists of another. Information is passed on like a handful of water, and after enough hands, there's no water left.'

He pushed open the door to the orthopaedic ward, the first I'd seen stickered with a no-guns sign. Ahead was the tail of a ward round, and I clung onto it, like an ignorant, servile medical student, grateful for a glimpse, for understanding.

'Things have changed. There used to be specialists doing open-heart and even brain surgery. These days, a so-called chest surgeon is someone who can insert a chest tube.'

That was a task not beyond my own capability once. Dr Ali himself had been invited for training in the UK, in Newcastle and Belfast, training that presumably would be put to immediate and vital use. But even with his medical licence, and reference letters in abundance, a UK visa was far from a given and he seemed reticent about discussing his chances.

'About 70 per cent of our patients here are victims of road traffic accidents,' explained Dr Ali, 'but this is old news; 20 per cent from shooting and such. This is growing fast too. Violence is like an infection, and it's not just gunshot wounds, it's psychological problems too, it's back pain, stomach pain, headaches. Family feuds can be settled using guns – we've seen people massacred at weddings. Never used to be like this, even five years ago. So much bloodshed. We doctors are the blood donors now. Even my son wants a toy gun. I tell him no. Ha! This cost me a bicycle! And even now, he makes guns out of paper.'

In the next bed an eleven-year-old boy looked up, his face distorted by fear, and not childish fear, something else, something murkier, more visceral. His mother, only a little taller than her son, reached for his hand.

Besmillah had been sent on an errand to the bazaar in the northern town of Maimana when a woman in a burqa detonated a bomb in a pressure cooker. The blast wave threw him into a nearby canal and he suffered a head injury and a broken femur. He was rushed to a clinic with no expert orthopaedic surgeon, where the only doctor applied external fixators, poorly, to adjoin the ends of his fractured bone. Dr Ali held up an X-ray film for me to examine: 'totally unnecessary' he grumbled, and I wondered if he was talking of the misaligned pins visible on the film, the incompetence, the lack of training, the bomb or the decades of war.

Dr Ali sighed.

'You wouldn't put him in traction in the UK, he'd miss three months of school. But we have different priorities here. We worry about infection. We worry he'll miss every day.'

When the bones failed to unite, Besmillah's mother took him to a mullah who proclaimed the boy to be cursed and responsible for his own pain and disability. It had taken him months to get to Mazar-e-Sharif after that, via the Red Crescent, and he waited now for further surgery and psychiatric evaluation. At night he woke, screaming and tearing at his bedclothes.

I offered his mother a seat, but she refused, crouching on the floor, gazing up at me, past the seat I wouldn't take either, speaking through a white veil drawn half over her face.

'He's not normal,' she said, speaking to the hospital floor. 'He screams and talks to himself. I pray his leg will heal, but I worry most about his mind.'

Before the bomb, her husband had become addicted to opium and she was forced to look after their six children alone. After Besmillah's injury, her other children had dropped out of school, or attended for only half the time. They were forced to work now, stitching together clothes to raise two dollars a day for food. And the violence rippled.

*

Hanging around Mazar was risky, and someone had already cut off one of the two padlocks we'd used to secure our room. It was time say goodbye to Sam, who was heading west to Iran. It was time to disengage – from the stress of conflict, from the checkpoints and chinooks. And it was time to exercise the power of my passport, and do what most people here could not. It was time to leave Afghanistan.

That evening, back in Uzbekistan and north of Termez, the sky was a pretty crimson and a tailwind buzzed in my wheels. I sang a little Motown as I rode. It's incredible the lengths the universe will go to in order to punish smugness. My pedal fell off. The oily bearings were lost somewhere in a 20-metre stretch of dust and I had to make do with the thread, my foot slipping off every turn, which created a rotating thought and sound: Fuck this. Clunk. Fuck this. Clunk. Fuck this. Clunk.

I acquiesced to the universe and set up camp on a hilltop. An hour after the sun set, a team of red flashes tore through the sky – not the thin race of a typical shooting star but the careering burn of something bigger, slower or closer, fragmenting. One piece popped white and stunning before the night returned to stillness. Through sheer luck and the blackness of the Uzbek sky, it was probably the only time in my life I will see a rare fireball meteor procession.

Towards Nukus, I passed cotton plantations like low clouds, then a gritty nothingness resumed, some biscuit-coloured plains, a distant factory, bluish against the dusk, then gone. A long stream of buses passed by, loaded with cotton pickers, fronted by police cars. There had been boycotts of Uzbek cotton because of forced labour and since 2012 children had been banned from working in the cotton fields, though the buses I saw were full of kids, faces planted onto the glass.

I crossed into Kazakhstan and cycled to Aktau, a port town on the Caspian. Before sunrise the sky held baubles of grey cloud and I made my way, in those delicious first moments of the day, past SUVs, testimony to the providence of oil, past statues of the writers and poets exiled by Stalin. A sign on a shop blinked redly: 'paradise'. The sun rose at last and the clouds burned lingerie-pink.

I'd found an American teacher, Paul, on the couchsurfing website and he was waiting for me by a stack of modern, well-to-do apartments. I followed him up some stairs to a flat that he shared with his Kyrgyz wife Jyldyz, some years younger than him, and her nephew, Islam, whose developmental delay was, too late, put down to deafness. Islam sat on Paul's lap, his hearing aid turned up for a bedtime story. Paul was teaching Islam to read, and he shone as a teacher. He was firm, but his challenges were kind ones, and he looked devoted in the task.

How had Paul pitched up here, teaching in western Kazakhstan?

'Long story, I guess. Raised Amish. Left school at thirteen. Ran off when I was eighteen and got excommunicated. Just jumped on a Greyhound one day, went up to Montana.'

'Why Montana?'

'Wanted to get as far from Pennsylvania as I could. Didn't want to be part of these ex-Amish communities neither. Needed a new life. It was hard growing up Amish. Didn't go to bars, wouldn't know how to conduct myself in 'em. I can't dance, don't know anything about sports. Kids probably thought I was a square or something.'

There was a fleeting lost look then, his eyes wide open but close-set, like fish eyes.

'Started logging at first. Helicopter logging. Hard work with that cotton-picking chopper buzzin' above ya. My family got my address, don't know how. Started receiving gospel verses in the post saying I'm gonna burn in hell, that I'm born Amish, gonna stay Amish. But then I got involved in this missionary project setting up schools in South Africa. I was in my element. Lived in a mud hut, two hundred metres from the Indian Ocean. I smelt wood smoke every morning.'

Paul looked dreamy, swept away by the memory of a life that must have been a half-decent surrogate for the nature-filled days of his childhood.

'I moved back to the US and learnt to be a teacher. The woman in charge was wary, what with my Amish background an' all. Turned out I was best in the class and got asked to speak at our graduation. I guess that's the Amish work ethic. When I was a kid, I didn't eat breakfast till the animals did. Dad tells you make eight holes in the ground, you better make eight holes or you get whipped.'

'Do you still go home?'

'Yeah, I go back every now and then, they single me out at the airport every time.'

Paul narrowed his eyes to approximate a border guard, did the voice.

'"Says here you been to Turkey …" Those guys are the devil's brother. I tell 'em, how many ex-Amish terrorists d'you know? But they always search me anyway.'

'You see friends? Family?'

'Yeah, I see friends sometimes, but they're hillbillies, you know? They think all black folk are on welfare. They don't know shit about the world. Bit like me back then. I knew a few black fellas growing up, me and my brother used to shingle their roofs, but didn't meet an Asian till I was eighteen. I walked right up to her and asked "Hey, are you from China?" Oh shit! She yelled at me! "I was born in the US! I'm as American as you!"'

He cackled at the memory.

Paul was making up for those years of seclusion. The TV in his flat was forever on the news channels, as if he needed to mainline the world and its mechanics. But his childhood, trapping rabbits and spearing sticklebacks with his brothers, still cast a shadow, and he disappeared for days at a time, bird-watching and hiking.

'How did you guys meet?' I pried when Paul was out of the room.

'On the internet,' said Jyldyz. 'But I didn't use my photo. I'm Asian! How shameful! I just stole a photo from some other girl.'

Paul came back in with two beers, handed me one.

'You guys travel together much?' I asked.

'Oh yeah! Every year we put the 238 nations and territories in the world in a random number generator to choose our holiday.'

'You're kidding?'

Paul grinned back at me.

'Where did you go last year?'

'Number 131.' I looked askance at Jyldyz.

'Liechtenstein!' she cried, throwing her arms up to lend it extra absurdity, before glaring comically at her husband. 'The hotel cost $100 a night. And mountains, ha! I have those in Kyrgyzstan!'

'The trekking was amazing though,' Paul chipped in.

'Liechtenstein!' Jyldyz screamed again.

'What about this year?' I asked.

'We drew Antigua.' He shot me a 'not-so-bad' look.

You had to love this democracy of theirs, their proportional representation, where the whole of the USA was on a level with Togo or Somalia. And I loved that Paul, from a community so insulated against the world, could play such roulette. What a beautiful over-reaction.

16

Consumed

The sun was yet to rise, but there was a faint, precursory glow as I made my way towards Baku, the capital of Azerbaijan. Beside the highway – tidy lawns, small trees clipped into spirals. The street sweepers were out already and clapped-out Ladas stood out like warts in a mob of gleaming hatchbacks. With daybreak, architectural wonders glimmered: the Heydar Aliyev Centre, with its contours like melting wax; the flame towers, skyscrapers, their windows blue with reflected sky. Baku felt too neat, too carefully primped, too show-city. And as for the tower blocks, I'd heard that many were empty, and that they served as bank accounts for a tight-knit band of elites.

And if wealth was professionally concentrated in Azerbaijan, so was power. I was reminded of the dynastic politics as I caught the gaze of Heydar Aliyev, the country's last president, taking up a large billboard. He was not smiling, but was not scowling either, his mouth a hyphen, promising more to come. And it did. When he died, his son Ilham dropped into his seat, commanding the oil barons, mobsters and cronies, silencing the press with a heavy hand.

West of Baku, rolls of grassland. A lunatic wind picked up dust. In villages, men sat playing Nard, a game like backgammon, then near Samaxi, fruit sellers swarmed around me. My first gift was an apple from an elderly woman. Her friend lumped a few more into my arms. Grapes arrived anonymously from over my shoulder before two men in suits lunged in, bearing tomatoes, shouting 'Present! Present!' Two ladies, who weren't about to let these men epitomise Azeri hospitality, pitched forward with aubergines and cucumbers and my arms were soon overflowing. Again, travelling had defied my crude sense of the world. We build places from gossip, half-truth and cliché, travel there and shake the foundations. I'd always known, of course, that there was much more to Azerbaijan than rapacious politicians and banged-up journalists – I was simply glad that now it was an armful of aubergines too.

Towards Ismailli, the odours of loam, mushrooms and woodsmoke pressed in. I'd missed these sylvan flavours, spent too long in the ice-clean air of the mountains, the gritty wind of the desert and steppe. I whooshed downhill, overtaking rattling cars that looked as if they might break up into a debris of bonnets and bumpers, like rockets descending from orbit at the wrong angle. Finally, near Qabala, I was released from the Great Caucasus mountains, into farmland that smelt of horses. Turkeys cavorted around heaps of hay. I foraged blackberries and nodded hello at old men tipsy on home-brewed wine. Forest gathered around me and sunlight dappled the road. I could ride a hundred miles a day here and still feel lazy. Even the farm dogs seemed to be wishing me good luck.

Balakan, the last town on my chosen finger of Azerbaijan, felt chillingly nationalistic, with its buildings and parks named in honour of the last president. With no particular desire to linger around the town's centrepiece, a flag pole many times higher than the tallest building, I crossed the Alazani River into Georgia. When I reached the capital, Tbilisi, the day was fizzling out and the Mtkvari River was liquid gold. In that fickle, conclusive way of travellers, I decided swiftly that Tbilisi, with its amphitheatre of surrounding mountains, had a good soul.

Happily then, my visa granted me up to a year here, a perk of Euro-friendly geopolitics and a relief after months of fretting about deadlines through the -stans. With time to spare, I'd arranged to cycle with a friend.

I'd known Olly from my student days in Liverpool. We built a friendship on a niche obsession for early nineties hip-hop (East Coast, naturally). We promoted gigs, fly-posted in the dead of night, and performed around the city with a hip-hop collective called The Punning Clan. But our thug days were behind us now, both of us pitching towards middle life, along with the rest of our crew, Benny Diction, Ro Jista, 2wo Toes, Louis Cypher and Tony Skank, who was now head of a large secondary school geography department.

Olly (aka 'MC Oscillate' between 1999 and 2003, thanks for asking) had flown into Tbilisi with his bicycle and together we began to plot a loop of Georgia. A loop seemed apt for a country outwardly all twists and curves. It wasn't just the roads and rivers wending this way and that, or the contour lines like fingerprints, towns too were marked in Georgia's shapely script. თბილისი means Tbilisi, but I'm sure you knew that.

I'd booked beds for us in a hostel called 'Why Not?' and as it happened, there were a number of good answers to choose from, like the tendency of the springs in the mattresses to make whirl-shaped scars over your spine and other pressure points. Our room was a grid of eighteen mattresses on the floor, and at night, when each held a body, it recalled a field hospital in Verdun. The other guests were young things, but theirs was a youth defiled. Pasty, dead-eyed and tattooed, like zombified blue cheese, they haunted the place in happy pants, hacking up phlegm and comparing spring-marks.

Olly had been given more time to appreciate all of this since he was already at the hostel when I arrived.

'*Why Not?*! Cos it's a pestilent slum.'

Maybe that was the hand of hip-hop: Olly was my most hyperbolic of friends.

'Come on, man. It's not *that* bad.'

My thoughts turned to the even less wholesome places I'd laid my head over the last few years: the buffalo shed in Egypt, the flea-infested animal skins on the floor of an Ethiopian barn, the noisy campsite-come-latrine beside a Chinese highway. Olly slept in a king-size bed with his girlfriend in London now.

He gazed around, agog.

'Look at that one!'

A particularly frail man shuffled across the room wrapped in a blanket. He began nibbling on a slice of bread like a sick squirrel that knows it's not going to survive the winter.

'Relax.'

'Steve look, I'm on holiday, I'm not working for an NGO, okay? This place is a humanitarian catastrophe.'

*

We woke late, stepped carefully over bodies and found lunch in a small café nearby. A man was slumped over a plastic table already crowded with empty beer bottles. Tim introduced himself, coherently by the third attempt. He'd been a rugby coach in his day, he said, though the sun was clearly setting over Tim. His face had turned the colour of wine, his eyes were lifeless and bloodshot, and I recognised his smell from the emergency department, the smell of booze metabolised, of a body eroding in daily tides of liquor.

Tim, like plenty of Georgians, enjoyed toasting. His arm had been wrapped around my shoulder for a minute or two when he raised his tipple again, a mind-bending moonshine called cha-cha, made from grapes, downed liberally throughout

Georgia, and considered the solution to many of life's frustrations, including acne, for which it was applied to the face.

Olly and I raised our glasses as Tim cast off on another of his rambling tributes.

'To the love of a woman! To my homeland!'

I made to drink, but Tim was not finished.

'To cha-cha! Wait! Wait! To world peace!'

'To world peace!' we yelled.

'To Benjamin Disraeli!' Tim yelled back, before adding, dolefully, 'The very, very best British prime minister.'

Olly and I swapped a querying look, but stood up and raised a glass anyway.

'To Benjamin Disraeli!'

We winced and spluttered. Tim knocked back the cha-cha like it was mineral water. He looked miserably around the room then, sunk forward, his elbow on the table and hand buttressing the ruddy flabs of his face.

'This used to be a decent place for working people. And now look. They're all vegans.'

*

It was raining hard as we left Tbilisi. Cars snarled through puddles, spreading mucky blankets that caught us broadside. Traffic bullied us towards an overflowing gutter. Olly was riding ahead, his head twitching this way and that as he tried to decipher the rules of Georgian roads. I waited patiently for him to learn that there weren't any.

After a few satellite towns, forest began beside us as the road switched over the Saguramo Range of Tbilisi National Park. A-frame houses could be glimpsed through the trees, half-forgotten and fairy-tale. Soviet mosaics appeared on walls beside the road depicting astronauts, aliens and glorious athletes. Weeds and weather had effaced the mosaics, and they were faded and cracked now, as if memories of a Soviet past had been left to ruin too.

No month could match the prettiness of October here, and every wave of land was forested in autumnal multicolour. The small acts of suicide had begun, and leaves, the colour of fox fur, were floating to the forest floor. Others were still green near the stem, but bevelled to blazing poker tips. The traffic was thinner now and as the forest breathed mist and the rain gave way to rusty sunlight, Olly was grinning as irrepressibly as me.

The Pankisi valley is not an area many people think to visit in Georgia. It lies down a dead-end road that coils into a rural community, home to an Islamic people called the Kist, whose origins could be traced back to neighbouring Chechnya. They numbered around nine thousand and were only noticed by the mainstream Georgian press when they popped out a fundamentalist, like Omar the Chechen, one of several top-level jihadis in the ranks of ISIS in Syria. I'd been invited to visit the Roddy Scott Foundation, an educational charity based here, named in memory of a young British journalist killed in the Second Chechen war.

Mountains appeared, a distant wall of peaks with a high border of snow. For hundreds of years, Chechens had journeyed over these mountains, on a route unmarked on maps, in dribbles at first, and in surges when Chechnya went to war with Russia in the nineties. Now, we transitioned absolutely. By the village of Duisi little of what I recognised of Georgia remained. Women came headscarfed down the road in tight, soft-speaking bunches, men with full beards watched us mutely, and an old lady unloaded a handful of chestnuts into my palms. Saudi oil had built the mosque and helped run the Arabic school, though education here was held by many to be tarnished by the ultraconservative doctrine of Wahhabism.

Organisations had moved in to try to busy the minds of the young boys, offering less bigoted dogma, but looking around at the minimal homes, many still wrecked from fighting between Georgian and Russian fighters, you could see how the lack of stuff might steer would-be shepherds to a fighter's life, which forecast adventure, money and celebrity, and might even help your family. One of the village's English teachers was rumoured to be married to an ISIS commander. She was the one with the new hatchback.

Near the Foundation, we met Cathy McLain, a canny educational psychologist from the US who ran a separate charity that supported disabled people and their families in the Kist community. Cathy's daughter, Lucy, had married Larry Page, one of the founders of Google, so fundraising was probably not something that required a great deal of her attention.

'When I first came here, the women heard an "expert" was in town. They lined the road with disabled kids in their arms like rag dolls, wanted to know which pill would fix cerebral palsy.'

'Where did you start?' I asked.

'We tried to help them not to feel sorry for their kids all the time. You have to encourage independence, work out what the kid's potential is. There's a lot of love, but it's not easy. The women say things like "We'll bottle feed him until he's 30!" Some hid their kids away, kept them in bed, even restrained them. Others thought cerebral palsy was infectious. People like us here, we're small, we don't have other agendas like some of the big NGOs. I don't compromise on my staff. We hire Georgians and they're *good*.'

We piled into a van for a home visit: Olly and me, Cathy, her husband Roy, a bullish problem-solver called Rezo, and a retired American family doctor. The house was comprised of two small rooms. By the far wall, a man looked at us from his bed, big black eyes, black tufts of hair, porcelain skin. His mouth was open, breaths rattling out. I noticed his nystagmus next – a constant, fast, involuntary dance of the eyes that, in some cases, speaks of brain damage.

The American doctor pulled back the sheets. The man's right arm was thin and folded unnaturally back, a contracture where the muscles and tendons had toughened over sixteen years of being bed-bound. When he was fourteen years old and full of beans he'd been at the town's annual horse race and festival. Firing guns was part of the festivities and he'd been accidentally shot in the head. The bullet was still in there, after sundering neurons, tearing through memories, senses and muscle control. He was blind and mute, paralysed in both his legs and his right arm. There was a broad hairless lump on his head where the bullet had shattered his skull.

Ideas were batted around between Cathy, the doctor and the man's family. Physical therapy, a radio to listen to. They sounded to me like marginal gains, but then where else do you start? I was reminded of a young man I'd once treated in London. He'd been beaten up one night and haemorrhaged into his skull. Surgeons had removed part of it to limit his brain damage. He, too, was paralysed, blind and nearly mute; he just wailed now and then. His mum had sat vigil, looking inviolable. Consultants delayed seeing him until the end of the ward round, and then we'd all huddle about his bed, a daily ritual that never contributed much, but forced us to see how one moment – a nanosecond, an inch, a snap judgement – can brutally warp a young life and all the lives of those bonded to his.

*

Since the 2008 conflict in South Ossetia, a large-scale Russian invasion of Georgia, by land, air and sea, on the pretext of 'peace enforcement', the region to our west was closed off. Russia defined the border with barbed wire fences and from time to time detained shepherds who ventured too close. So we backtracked and from the tidy cathedral town of Mtskheta turned west to Gori, a town occupied briefly by the Russians in 2008 and still ranked with hastily put-up houses for refugees. Gori was Stalin's birthplace and a museum commemorated this fact. Our tour guide was an ice queen for whom tourists were a burden, if not an insult to the memory of Stalin himself. She raced us through his affairs.

'This picture of Stalin. This stuff of Stalin. This presents for Stalin. Next room. More stuff. Stalin very popular, you can see. Okay finish. Questions!'

She yelled this at point-blank range and in a manner that suggested that asking one would precede our forced exile to a gulag. Nobody dared.

The big finale was still to come: Stalin's death mask, presented on a plinth, which I think the tour guide took home with her each night and wore as she slept. The purges, the banishments, the deal with Hitler were not included in the tour, though a small corridor on the ground floor did treat them as a footnote.*

We stopped at the village of Dzevri next, where friends of Cathy had invited us to a *supra*. A *supra* is a full-bodied Georgian tradition, sometimes described as a feast but really a kind of concerted celebration of all the things Georgians love to indulge in: food, drink, memory, love of country and song.

It was taking place in a spare, two-level concrete building dating from Georgia's Soviet days, below mountainsides sunk with caves. We were guided upstairs to a room containing a long table crowded with wide men. *Supra* means 'table cloth'

* Stalin wasn't all good.

and a good *supra* is so bounteous, you shouldn't be able to see one for food and drink. This had been achieved bar a small area in the centre of the table where a suckling pig was now thunked, having been spirited, shiny with grease, from a village fête with the blessing of a priest. Small cheese-laden pastries, *khachapuri*, were dispersed. Wine was in ten-litre plastic flagons, home-brewed, for slugging, not sipping.

At the head of the table sat the Georgian toastmaster, the *tamada*: derived from two words – 'tam' (all, everybody) and 'ata' (father). Our daddy had a drinking problem. He was a fleshy, salt-and-pepper hardman, and like many of the Georgian men I'd met so far, he wasn't much of a smiler. A regiment of big bellies and rugby noses lined each side of the table, all the way to a knot of young, furtive men at the other end. There was little to distinguish it from a scene in *The Lord of the Rings*.

Olly sat to one side of me, and the only woman at the table to the other, translating for us.

The toasts began in the expected order.

'To God.' We drank, our religion demanded it.

'To all of Georgia, including South Ossetia and Abkhazia,' boomed the *tamada*.

There were murmurs of assent at the mention of territories effectively now Russian. We drank. A toast was made to Olly and me, Georgia and Britain declared eternal friends.

It was a more open affair now, men stood and spoke as they felt, told jokes or sang. We lauded women and children with more toasts. A man opposite took his go.

'The Angel of Death comes for Miko, your time has come, he says. Wait, wait, says Miko. I have some wine, drink with me. The Angel of Death drank wine with Miko. Eventually the Angel of Death says, you go on ahead, I'll catch you up.'

The men thumped the table, juddering with laughter. A man to my left stood up.

'Under the Soviets we were told to stop producing wine!'

There were murmurs at the reference to Gorbachev's alcohol reforms, which had decimated Georgia's wine industry.

'But how can I stop producing wine when I am always drunk!'

Another thunder, fists clouting wood.

The door swung open and three men moseyed in with an accordion, drum and clarinet. We anticipated a slow, organic sound, but there was an explosion: the clarinet was out of tune, the drum smacked hectically and without any cadence, the accordion was wild and blasting. The din died 30 seconds later; the men looked proud of themselves and left. At the prompt of music, a heavy man pulled himself up, took a breath and filled the room with his lusty baritone. The woman beside me explained 'They sing about a girl, her eyes the colour of the sky …'

As his song melted into a thoughtful peace, the *tamada* stood again. In instant deference, we looked to him, hushed.

'To our parents. We are stronger with our parents alive. God bless those who have died.'

A silence returned, as if ghosts were gathering watchfully above us. The oldest man at the table, wearing an old-fashioned tunic, clipped with large bullet casings and a small sword, stood up next.

'I knew your fathers. I knew your grandfathers!' he reminded everyone.

A man asked, 'Does the sword still have action?' The innuendo got another peel of laughter.

The *tamada* stood again and looked at me. 'And let us toast Tbilisi! Who drew with Tottenham Hotspur in 1973! Do you remember?!'

'No!'

There was excitement. My ignorance must be remedied.

I didn't know the score line of a UEFA Cup football match played seven years before I was born, involving two teams I didn't support, playing a sport I didn't care about. I toasted the one-all draw as I'd toasted Benjamin Disraeli; why not?

A bear-like man slapped his paw on my spine and said something I was too drunk to decode. The *tamada* surveyed a job well done. The piglet was picked clean, everyone was bonded, Georgia had been united with lost territories, our dead parents were proud of us, everyone and everyone's wife had been celebrated and, vitally, we were all now completely twatted. '*Gaumarjos!*' We yelled. 'Victory!'

*

'Hung-over to fuck' was how Olly correctly summed up our predicament the next morning, a situation not helped by *khashi*, traditionally a cure for hangovers in Georgia. *Khashi* is a broth, environmental-catastrophe yellow, in which gristly and hairy things bob around … loops of bowel, some lung, probably. I'd managed one spoonful: a rancid attack of salt, butter, oil and flesh that made my eyes gush. We clambered onto our bikes, uncured.

A 600-metre ascent in hairpins eventually put paid to our headaches. Around the Shaori reservoir, the eerie mood of autumn seemed to echo in the villages: tumbledown homes of planks and sheet metal, overgrown orchards, quick-to-anger dogs. We were heading into the northern mountainous region of Svaneti now simply because it was rumoured to be enormously beautiful. The atmospheric dawn over Lentekhi was already providing some sense of this. Odd coppery trees burned in an army of evergreens, clouds looked snagged in treetops and great clefts in the mountains produced crashing waterfalls.

We climbed. The steepest bits were also the rockiest, and

so steep that my front wheel left the ground in an involuntary wheelie, like the rearing of a stubborn horse. We were coming into Ushguli now, a community of four villages said to be among the highest inhabited settlements in Europe, though defining inhabited, settlement and Europe has obvious snags. It is backdropped by a massif, the Bezingi Wall, a 12-kilometre-long ridge including Shkhara, Georgia's highest peak. The villages were scattered with medieval-looking square-shaped stone towers, *koshki*, doorless at ground level, suggesting an embattled past. The Svan people had a reputation in Georgia. 'Don't drink with the Svan,' I'd been warned. 'It's the altitude. It makes them want to fight. It turns them to a kind of wood.'

We went completely noticed in Ushguli, thanks to Olly's uncanny ability to attract every dog in town. There were big bruisers and toy darlings, hale and rabid, all snarling and barking at his wheels. Three leapt up to his bike, their paws on his rear panniers, and he careened around, wailing 'Get away! I'm not a plaything!' The drama was even more arresting via my camera's viewfinder. Flapping his arms shed a few vicious ones but he kept a cute shaggy specimen. Olly looked at it, half-lovingly. 'Yoooooou festering fuck-bunion,' he said, to which the dog pawed at him in appreciation and then trotted happily behind his wheels.

We lost our new friend on the downhill, passing smoky-coloured horses pondering the hillsides. Beyond Mestia, the rain brought a sparkle, and wet persimmons jewelled the trees. The forest was mostly pine now, but winter was not far away and there were a few elm and ash, standing naked, reminding me of how Nabokov compared trees in winter to the nervous systems of giants.

After we looped back to Tbilisi, Olly flew home and I moved into a large guesthouse run by a pathologist. She was kind to me, kooky, and every bit her profession. By evening her black bob was bent over a microscope, her hand flicking away ash as she chain-smoked over slides. Her daughter, Mariam, hung around the hostel at times. She was a red-headed film-maker, partnered with a British man, also a film-maker, called Nik. I eavesdropped on them shamelessly, like a travel writer should. They were planning to make a documentary on a hospital for patients with tuberculosis in the hills south of Tbilisi. We soon got chatting and I was invited to join the crew.

As we drove to Abastumani, Mariam gave me her take on Georgia today.

'Politics is fucked up here! We've just had an election. Before the vote, people in Freedom Square were encouraged to write down what they wanted on pieces of paper. People have no self-respect. They wrote down things like "washing machine". Fuck! The new president was elected with Russian money. And do you know what was debated in parliament next? The right of Georgian officials to wear full regalia to international summits, which will be fucking embarrassing, and the banning of condoms that have ribs for pleasure.'

Towns grew further apart, the road winding through the shadows of hills. Dashes of snow filled the gaps between pines. The bus pulled off the road and Abastumani came into view.

The clearest quality of the estate was a fading grandeur. There were roughened sculptures, a small chipped convent, a ruined fish pond. Cows loped in front of the hulking main hospital building. The walls shed paint, creating what sometimes looked like a map of an unknown earth. The building was like some great wounded animal now, a beached whale perhaps, rotting to its brick-and-mortar bones. A place of decay, where patients with tuberculosis still came seeking cure.

In 1890 Grand Duke Georgy Alexandrovich, a Romanov and younger brother of Russia's last tsar, Nicholas II, developed a chesty cough. His mother sent him to Japan first, hoping 'that it will pass with a change of air'. TB has long been considered treatable by exile, and the coast, desert and mountains have all, at different times, been considered beneficial for consumptives. Keats heeded medical advice and moved to Rome, Robert Louis Stevenson went to the Pacific, and Chopin headed to the Mediterranean. Japan didn't help the grand duke however, he lost weight, found it difficult to breathe and grew melancholy, earning himself the nickname 'weeping willow'. Young Willow moved to the sanatorium at Abastumani, and then one day, to break the monotony, he took a ride on his motorcycle. A peasant woman watched him collapse in the road. Blood oozed from his mouth as the 28-year-old lost a battle for breath. Rumour has it, the peasant held the Grand Duke in her arms as he died.

The story of TB in the modern era is one of innovation and clever science, but also one of complacency, bias and cock-up. Throughout most of the nineteenth century, many patients with tuberculosis in London turned to Poor Law institutions and could expect limited sympathy. I have my favourites in the panoply of causes that have at one time or another been linked to TB: Irishness (well, of course); 'A mournful tendency of the soul' (seems a little unfair to judge this a cause and not an effect of TB); impure air (correct); prominent shoulders (nope); depraved character; the sorrowful passions of young lovers. Famously, TB in this era had something of a romantic cult; it was widely believed to be a disease of creatives, and sometimes thought to make people more interesting, sensitive, passionate and soulful. Deaths from TB were portrayed as nearly beatific.

Early attempts at treating the disease were not evidence-based of course – this was back when medicine was mostly human experimentation and, as Lewis Thomas notes, 'based on nothing but trial and error, and usually in that sequence'.

Bygone remedies included the ancient Roman habit of drinking elephant blood. People ate wolves' livers and bathed in human urine (which presumably worked by ensuring that, for a short time, a bloody cough was not the worst thing about your life). English and French kings and queens touched people with the swollen neck glands characteristic of TB to cure them, hence 'the royal touch'.

In the nineteenth century, on noticing that TB was linked to urban areas, it made sense to remove sufferers from this environment. Some suggested that the thinner air of mountains would allow the blood to circulate more freely and by the end of the century, sanatoriums for consumptives were all the rage, Abastumani included. While 'triple therapy' nowadays relates to the three drugs commonly used together to treat TB, in the nineteenth century it meant fresh air, a good diet and carefully planned sleeping hours. There were no drugs against TB until 1944 when streptomycin fronted the revolution.

Though 90 per cent of those infected with TB never suffer a single consequence, it is not easily defeated once it digs in. Tuberculosis is a mycobacterium, the ponderous tank of the microbial army. A bug like *E. Coli* takes twenty minutes to divide, while *Mycobacterium tuberculosis* takes nearly 24 hours, but although the bacteria might grow slowly, they have an impervious wall of glycolipids to keep them in business. The discovery of the specific bacteria by Robert Koch in 1882 was during a time in which a seventh of his fellow Germans died of TB, though it was already on the downturn through the segregation of patients and improved social and public health. And yet we remain far from eliminating the infection – TB kills more people today than when Koch made his discovery.

By the late 1970s, the medical establishment, regarding the record of infectious disease, thought it had it made. Smallpox: defeated. TB spiralling, surely, towards oblivion (in the richer world at least), following its sibling leprosy down the plughole. The new drugs against malaria were full of potential. There had been no serious new nasties for decades. And then the 1980s came, and the world began to wobble. HIV crashed the party first. Resistance to anti-TB drugs, other antibiotics and to anti-malarials accelerated. Superbugs began to plague hospitals. There began large-scale denial in face of the facts.*

In contrast to the nobles and celebrities of earlier years, the patients today in Abastumani were at the other social extreme. Exiles, end-of-the-liners, ex-cons. Some were not on treatment at all and many would die here. They came because they were broke or their treatment had failed. They came with strains of TB resistant to antibiotics. A few had been 'thieves-in-law': men within an organised crime network in the USSR, one-time leaders of prison groups in forced labour camps. After Stalin's death, around eight million inmates were released from gulags. Thieves-in-law were drawn from many corners of the Soviet Union, but at least one third were Georgian.

In the 1990s, after Georgia's civil war, organised crime was rampant and the state was frail and bankrupt. Mikheil Saakashvili, inaugurated as president in 2004, began using tactics similar to those used by the Italians when they clamped down on the mafia. He was ruthless. There were mass arrests and confiscations of property, and officials turned a blind eye if the police were heavy-handed. Soon videos surfaced of gangsters being tortured. Sentencing in the courts was designed to send

---

* HIV (which increases the risk of developing TB by 20-fold) in particular suffered from this, with prominent South African politicians refuting its role in AIDS. But to claim that HIV was raging across Africa fed the stereotype of a dark, dangerous, desirous continent. Once that was unavoidable, it became difficult, politically, to ask why it was spreading so fast across Africa, unless your answer was poverty, which had a role, but was only part of the answer.

a message, and as prison numbers increased, the cells quickly became cramped. At that point, Georgia had a higher prison population per capita than any country in Europe, higher than Russia too. TB thrives in overcrowded prisons.

The patients inside Abastumani were mostly men. They wandered the corridors, pulling on cigarettes and wheeling around, visibly drunk. I watched how they talked with each other, and which of them approached us. There was a strong sense of hierarchy here, borrowed, perhaps, from prison life. I heard wet coughs from behind closed doors. The film crew held cameras and a boom, and we were a curiosity, but most patients stayed their ground and we pulled in only the cocksure, and the drunkest men. Uncharacteristically, then, a young woman approached us in luminous pink leggings. Her hair fell thinly around ringed, hazel eyes and a small mouth with bad, pointed teeth. She looked about twenty. She spoke in Georgian to Mariam who gave Nik a thumbs-up. She'd talk with us.

We followed her, the men looking insulted that the pecking order hadn't been respected. Her room was chilled by an open window that admitted the odd puff of snow, like a machine for the stage. An electric heater worked against the natural ventilation. I looked out: the pines outside were frosted today, a solitary dog wandered among them, sniffing for food.

She sat on her bed with her arms crossed in front of her and I watched her eyes wet with tears as she smiled in spite of herself, struggling to keep her emotions at bay. Mariam sat down beside her.

'Are you okay? Where are you from?'

Khatuna had grown up in a large Muslim family in Adjara, a region in the south-west of Georgia. Her father had lost his fingers in an industrial accident and she'd been forced to earn money young. She was married in Turkey, beaten by her husband, got divorced, grew broke, got deported. I wondered if she'd been involved in sex work because of how certain details – the poverty, deportation, rejection by her family – orbited the unsaid. None of us imposed the hurt or embarrassment the question would have caused.

Nobody in her family knew who'd contracted TB first, but her brother, sister and father were all soon in its grip. For her, six months of treatment had failed. She'd been in Abastumani for a month now; it had been tough to source the TB medication from home and she'd suffered side effects.

'I was worried about coming here, I heard you could be raped, that it was full of prisoners. I see people dying here. What's the point of the pills if I'm next?'

Past inspections of Abastumani had revealed widespread problems: outdated tests, mingling of patients with resistant strains, unsuitable drugs. There was no central heating and the temperature in winter rarely got above zero. It was a place of myth and rumour, warped facts and fictions, wild hopes and whispered despair. Some said that the doctors were infected; others said that the air was so pure you didn't need drugs here.

We moved on and collected in Giorgi's room, further down the corridor. Giorgi could have been around forty. His beard overgrew a strong jaw, his eyes were close, his words fell slowly and, it seemed to me, with an air of defeat. In 2013 he was released from a four-year term at Ortachala prison, a place so rife with both abuse and TB that it has since been shut down.*

'The conditions were bad. There were beatings. The staff threw cold water on you. I saw torture. A lot of people died from TB there. Nobody cared if you were sick. Everyone was sick with something. You had to be rich or literally dying on the ground for anyone to care. I saw guys cut themselves to get medical attention.'

---

* Abastumani itself was demolished in 2019.

'When did you get sick?' I asked.

'Inside. I had fevers and a bad cough. Then I lost the vision in this eye, but nobody did anything, they only told me it was TB after I got out. I'm kind of a problem child,' he said with a smirk. 'I drink. In prison too. If you can get taken to the sanatorium you can steal the medical alcohol.'

'And after you got out?'

'The TB had affected my nervous system and it was resistant. I took different drugs for eight months but they say there is still TB in my chest. They say the drugs won't work for me. But the air is good here.'

A couple of men stumbled into his room and grinned drunkenly at us. Giorgi asked them to leave. As they did so, he lit a cigarette. Ash toppled, the dust whipped away by that good Abastumani air.

'If I had money, things would be different. I know that. I did wrong and paid a price, but the price I paid was too great. I take 22 tablets a day and I get injections, but I see people on the same treatment as me dying.'

There were others, similarly hopeless, reedy patients whose chaotic lifestyles meant that they didn't always finish treatment courses and so rebounded between TB facilities spreading more resistant strains. Some patients had been thrown out of Abastumani ten times or more for drinking.

*

'Societies are judged on how they treat their most vulnerable.' The quote is often misattributed to Mahatma Gandhi (who may have said something similar, but in reference to animals, not people) and it has been adapted variously over the decades. Churchill turned the 'most vulnerable' to 'prisoners', others have shifted the spotlight on to the impoverished, the elderly, the sick, the handicapped and, if you're an American Republican, the unborn. As we drove away from Abastumani, I wondered whether, on some level, societies blame the most vulnerable for being sick, whether we feel it's their due. Not only in the way the tabloids encourage us to – life choices causing illnesses that cost the taxpayer – but for a deeper, more impulsive, subconscious reason. Historically, this feels true. I thought about how the ill have always been demonised and ostracised, from lepers to those sealed away and chained up in asylums; and from Typhoid Mary to the first of the AIDS patients, the sharp end of a 'gay plague', as it was then, with its allusions of divine retribution. We have always attached moralistic significance to disease.

In *Illness as Metaphor* Susan Sontag compared the punitive or sentimental fantasies we concoct about illness to how we stereotype national character: 'There is a link between imagining disease and imagining foreignness. It lies perhaps in the very concept of wrong, which is archaically identical with the non-us, the alien.'

There are abundant examples through history of foreigners themselves being unfairly blamed for disease. The Irish were held responsible for spreading cholera in nineteenth-century London, Jews for introducing plague into fourteenth-century Europe. The evolving etymology of syphilis is edifying. The disease was definitely someone's fault, but whose? Writing in 1588 Juan Almanar described syphilis as 'the disease known among Italians as Gallicus, that is to say, the French disease'. The French called it 'The evil of Naples'. Lopez de Villalobos, a Spanish writer, called it 'the pestilence of Egypt'. Jewish communities expelled from Spain called it 'the Spanish disease'. The Australian Natives Association in the 1920s, which had serious concerns about the spread of communism, dubbed syphilis 'the red plague'.

It's easy to assume that judging sick people as morally corrupt is an antiquated way of thinking but, like foreigners, the sick will always be soft targets and ready scapegoats. No patient is without agency, but no patient is simply feckless either. They don't come to Abastumani or St Thomas' emergency department simply because of a lifestyle choice (every death, under those terms, part-suicide). Many of the patients I'd treated in London had ruined their health – but it is deplorably shallow to ignore the reasons why.

It is tempting though. For the healthy in particular, it's comforting to discount the undercurrents of illness. I suspect that this is because we want to deserve good health: we like to take it for a virtue or a prize, in the same way that we like to think of other fortunes as profoundly personal and nothing to do with chance, or circumstance, or the fluffy mattress of privilege. We're all guilty of this. And I'm sorry to say it, but that means you are, too.

17

Flow

Bish, bash, bosh, I thought, leaning over a map of Europe. I'd crossed from Georgia to Turkey, Turkey to Bulgaria, and I was in a hostel in Varna now, a city and holiday resort on the Black Sea coast. Europe, the final continent, was unfolded before me, spider-webbed in roads. The map's key told of service stations, chalets, roman ruins, even speed cameras. Which way home? I was blessed with choice. I recalled my map of Uzbekistan, which had been so lacking in detail that it sometimes seemed like the cartographer had died abruptly before the job was done. I'd felt strangely full in the desert though, full with the intensity of the place, the soothing stillness, the ringing space. I wondered now if the tangled roads and abundant towns and cities ahead would make for a less restful ride.

Once again, Europe was in the grip of winter, but I imagined this crossing would still be smoother than my first. I wasn't that fleshy novice any more. I'd changed, surely, yet I had no yardstick, no friend who could assert who I'd become. The man in the mirror looked trim, athletic, even, in the right light, but he'd been visibly affected by a thousand nights rough camping and a less than social lifestyle. There was a shiftiness about him, a sort of grim acquiescence in his eyes. He looked like he might abduct your grandmother and ransom her for Wotsits and scrumpy. For his sake, I needed to be kind to myself on the home stretch. I would do away with hills, with maps, and the paralysing problem of choice. I would breeze home on a route a pensioner could reasonably master, and often did. I would ride the Danube bike path.

The Danube drains nineteen European countries and flows through ten of them, more than any other river in the world. Out here, in the east, it also played dividing line between Romania and Bulgaria, and the first few days passed in flat, fast miles and quick coffees on both banks and in both nations. There were little acts of kindness, and I cherished them more than ever, for they sustained the great theme of my journey to the end. A service station manager in Romania bought me a sandwich. A Bulgarian hotelier gave me a night for free and a bag of doughnuts. By the new year, that January light had spread over everything, pale, unmoving, and when ice gathered in the early mornings, the world seemed ossified, and the trees became ghoulish, reaching things.

The Danube and I wended through Serbia, Hungary, the frozen wetlands of Slovakia, and into Austria. In the hopeful and dazed manner of homecomings, it all felt so typical, so intimately European – from the Slovak villages that smelt of baked bread, to the hot wine and clove-scented markets of Budapest, to the young women in Vienna carrying tote bags printed with 'Facebook is not Franz Kafka'.

There were something like 71 million signs on the Austrian leg of the Danube path, three or four different maps at each information point, and they seemed to roll around every kilometre. Zoom-ins, large scale, different angles, extensive keys, historical titbits, elaborate descriptions with photos of wildlife and local guesthouses. There were maps that told you the location of other maps. There were altitude graphs, almost completely flat lines, in case you suffered from some terminal brain disorder and had forgotten you were following the course of a large river and its floodplain. It reminded me of the short story by Borges in which a town is so dedicated to making a detailed map that they eventually make one bigger than the town itself.

I crossed from Austria into Germany at Passau, where two more rivers, the Inn and the Ilz joined the Danube and where a young scamp called Adolf Hitler was reputed to have fallen into the icy waters while playing tag with his friends. His life was saved by a passing priest. Whoops.

Beyond Passau, I began to wonder if this still counted as cycle touring. More pertinently, I was now on an extensive tour of German bakeries. Cycling had become somewhat incidental, a means to an end, and the end was strudel. I'd imagined coming home a toned, serious champion; more likely, I'd return a very happy man, with bits of pastry on my face and conspicuous moobs.

The tarmac became sheeted in snow again towards Winzer. One morning the rising sun anointed pines overhanging the bike path. Overnight, fresh snow had stacked finely on the boughs, no breeze to upset it. The moment the trees were touched by the new day, snow sailed down in puffs and flecks, so that the whole path was at once prettified, a blizzard beneath a solid blue sky. It was a spectacle so dazzling and so life-affirming that there is probably a specific word for such a phenomenon in German. And that word is probably Shruntabintafrakan.

The Danube bike path would be rammed with cyclists in the summer months, though I saw none now. 'Camping' signs in Germany to lure the summer bikers, crowned with snow and lustrous with ice, looked more like warnings not to. *Arschkalt* they say in German: *arse cold*.

As I couchsurfed across Europe, talk with my hosts often strayed to what the papers were calling the 'refugee crisis', a crisis perpetrated, depending on who you talked to, either by the refugees themselves, or by what compelled them: the conflict and globalised inequities behind their existence. The previous summer's exodus was said to be the largest movement of people on the continent since the Second World War. I saw men in Munich distributing flyers from PEGIDA (the acronym, translated, stands for Patriotic Europeans Against the Islamisation of the West). Left-wingers had been galvanised too and back home, 'Choose Love' became a slogan for the resistance. Several of my hosts in Austria and Germany had assisted at the refugee and migrant camps in Calais and Dunkirk, others had hosted Syrian families and Afghani men. The headlines in British newspapers evoked tidal waves, rivers, swamps and other water-features of displaced people, while the *Daily Mail* indulged in more toxic metaphors: migrants swarmed, flocked, besieged. If anyone could take the shine off the prospect of coming home, the British gutter press could.

*

I left the Danube at Regensburg and turned north, heading for a small and generically German town called Hövelhof in western Germany. It was a place that would have had no appeal at all, were it not for one of its residents, Heinz Stücke, the world's most dedicated cycling vagabond.

The stats of Heinz's ride sound implausible, and when I recite them, people tend to scrunch up their faces and look at me for a time, silenced and baffled. Heinz is reputed to have cycled 650,000 kilometres all told: a distance equivalent to sixteen times around the equator. Two years ago, he'd finally returned to Hövelhof, and by then, he'd spent 51 years – 51! – cycling around the world.

After dead ends and bouts of unanswered phone calls and emails, I'd finally managed to get a phone number for Heinz and he'd invited me to stay for a couple of days. Heinz was the ghost I'd been tailing for some time. Six years ago, in Nice, I'd stayed with a woman who'd once seen a small man sitting on a Tokyo side street next to an old bicycle, flogging leaflets. She'd bought one from him, the man's life story abbreviated to a few glossy pages of words and photos. Heinz had spent decades selling these leaflets in major cities to bankroll his travels and that night in Nice, I stayed up reading the leaflet, captivated by the near misses: when Zimbabwean rebels shot him in the foot and stripped him to his underwear; when he'd been stung almost to death by bees in Mozambique. His moments of good fortune too: Heinz had been presented to Emperor Haile Selassie, who'd given him $500. In Thailand, a plane flew over his head and made an emergency landing on the road. Heinz's stories were rarely less than awesome. Passing through Hong Kong, I'd called in to Mr Lee's Flying Ball Bicycle shop, open since 1940. Mr Lee had employed Europeans and Hongkongers as bike mechanics, and they had wheel-building contests, eastern-style versus western-style, like they were martial arts techniques. It was Heinz's favourite stop in the city. Mr Lee sold Heinz's leaflets too and he was one of the few people he counted as a friend. He let Heinz sleep in the shop's store room whenever he was passing though.

Heinz's directions to the first home he'd ever owned were exact, as you'd expect. Trees and benches and each possible landmark were included. I counted out four trees and found a white bungalow on the edge of a school compound, a lamplit globe in the window.

Heinz threw open the door. 'Where have you been, I've been waiting for you! Why you ride in this rain!'

'What! You never cycled in the rain, Heinz?!'

'Well of course, sometimes you have to!'

He grabbed my bike and together we wrenched it into the hallway.

'My God it's filthy. Wait here.'

He returned with a toilet brush and began attacking my chain.

'There's no excuse for a muddy bike! There you go. You're one kilogram lighter now.'

Heinz was short and I was instantly pleased that the legs that had turned so much to criss-cross the globe were diminutive ones. A great bald forehead divorced two grey refuges of hair. His brow was substantial, occasionally lending him a ruthless look, but his eyes were a lively, cloudy blue. There was a slight resemblance to the presenter Clive Anderson.

His bungalow had been donated by the town in homage. Five rooms came off the hallway, and the walls of the largest room were covered in blown-up photos from his travels: cycling through an early morning of backlit baobabs in Madagascar; lolling in some rare shade beneath an acacia; cycling beside a rickety bus in Vietnam. He noticed my interest.

'Incredible, those buses ran on wood! Look, you can see: wood burners at the back of them! Ahh, they don't exist any more.'

I grinned at an onrushing wall of sand.

'Those sandstorms in Mauritania were horrible, but not too long. Now this one in Niger, it went for ten days!'

I'd seen a few of these before, like the one of Heinz's darker side: a self-portrait in Iran after a car had smashed into him. His face is laced with fresh scars, and he's glaring.

My favourite photo of Heinz, however, was not on display, one taken in 1960, two years before he left home. Looking younger than his twenty years, he straddles a bicycle too big for him, in short shorts, with a quiff of hair. He stares past the camera with a strange intensity. You might consider Heinz adrift, many have, but in that particular snapshot, he has a destination in mind. I wondered now whether he'd made it.

We sat in his kitchen and for the next ten minutes I barely

got a word in. I'd heard this of Heinz. In 2006, when he was riding through Portsmouth in the UK, his bike was stolen for the sixth time. The British papers were livid: 'Theft puts brakes on round-the-world cyclist.' While Heinz made public appeals for his bike to be returned, he stayed with a local man who'd offered to help and who was quoted in the *Daily Mail*: 'I've never met a man who talks so much. I'm at my wits' end. We can't switch him off!'

When you imagine the kind of person who would up and cycle for 51 years, perhaps you visualise the venturesome stereotype: game-faced, aloof, thousand-yard stare. What you don't imagine is a jaunty chatterbox like Heinz. I didn't think that his place would be so cluttered either. On the wall of the living room hung his framed Guinness World Record certificate (he was the most travelled man in the world at one point in the 1990s, before a whole fleet of obsessives began jumping on and off planes to collect outlandish passport stamps), and nearby, stacks of his hardback *Home Is Elsewhere*. There were more books, a couple by James Michener, one by Nietzsche, and a 1974 copy of the *Sunday Times* magazine with an article about Heinz, 'The man who wanted to see it all'. Filing cabinets were crammed with tens of thousands of slides, shelves carried hundreds of maps and journals. This had been his task for the last year. Heinz was trying to catalogue the lot, though he struggled with computers and had no email address.

When Heinz arrived back to Hövelhof, the town had grown to three times the size it was when he'd known it. Some of his old school friends were still there and had barely left. He hadn't seen them in half a century.

'Was it emotional?' I asked. Weeks away from London, I found myself tearing up at the prospect almost daily.

'No, no, no. It was nice. Mayor was there, we had a small celebration.'

There were three globes in the living room, one of which looked like a team of toddlers had attacked it with coloured crayons – the scribbles described his movements by bicycle. Heinz was chock-full of facts and figures and, in a few seconds, he reeled off the land masses in square kilometres of all the continents, boastfully. Heinz wasn't happy with the map I'd knocked up of my own adventure. It was only a rough interpretation of where I'd been, way below his standards.

'But what's this? Xian? No, no, no, that's much too far east. And here? You're way off.'

In the next room stood his first bicycle, painted with the names of countries and cities he had, at one stage, rolled through. He had a stack of 144 small strips of rubber, each cut from a tyre he'd worn out. There was an anaconda skin on the wall and a pile of watches and spanners he'd found on the side of the road – waste aggravated him and on occasion he'd tied knots in inner tubes to solve punctures. For years, he'd posted this stuff to his sister in Germany or to a place he called 'the bunker', a dusky room in Paris, lent to him by a friend and a brief base camp from time to time. An array of hats were displayed on the table too.

'I love this one, from Australia, see, keeps the sun away and I can mop myself with it!'

He did so.

'I swear, I sweat more than any person alive!'

Heinz had actually broken a bike frame in this manner: the sweat dripping from his nose had slowly corroded the cross tube, like a human stalactite. Years of sun exposure had given him cataracts. He had a recurring nightmare, he told me: someone steals his bike and slides, he has to chase them down, running madly, leaping over rivers, always in vain.

A collector, then. Impossible not to connect the photos, hats and chopped-up tyres to the miles, countries and lines on a globe. There was an obsessiveness in evidence in his journals too, where he'd meticulously recorded the days he spent where, the number of visits, and other remarks, all of which he allowed me to skim through. On one day in 1973 he'd recorded having six beers with a British biker. A map was spread out on the bed: 51 Christmas days had been marked by location.

I didn't need to confirm my thoughts about Heinz, but I did anyway.

'Heinz, is this an obsession?' He absorbed the question blankly. 'It's what I do,' he said. 'It's all I do.'

He brightened.

'I cycle 12,000 km per year, minimum!'

I wondered how goals like this validated his travels, and which ambitions were, to him, the worthiest. I'd paraded my own goal of cycling the length of six continents once, impressed by my own determination. I used to take pride in how long I'd been on the road in a way that was embarrassing to me now. What was meaningful about notching up miles? If such things bespoke qualities like perseverance then perhaps that was true for the first couple of years, but speaking about my journey felt more confessional now. *Hello, my name is Stephen, and I'm a monomaniac.*

To get Heinz wound up was easy. There was a reactionary in him. That evening he bemoaned the demise of short-wave radio, the frustrations of applying for visas on the internet, and the latest model of a North Face tent that let water accumulate between the layers. But he'd often lose his train of thought and the next moment he'd be demonstrating how to cope with 100 per cent humidity in the Amazon, stretching himself out on the kitchen table and fanning himself down with a wrapped-up towel.

'You see! Like this! This is how I do! Like this!'

Heinz made me some dinner, a spread of bread, sausages and cheese. 'But first, a warmer!' he said, pouring me a shot of a dark brown herbal spirit, Schierker Feuerstein. Afterwards he served up tea in massive cups, infused with camomile ('like the Indians do!') and we discussed for ten minutes the different teas of the world, from the mint teas of Morocco to the milky chais of the East.

'People want superlatives,' Heinz complained. 'They want to know my toughest, highest, best ... places. I don't know.' But he hated questions about his personal relationships even more. He didn't make it to the funeral of either of his parents and couldn't have known of their deaths for weeks, maybe months, afterwards. Heinz had had a girlfriend for eight years, a Belarusian called Zoya ('Zoya The Destroyer' – with a wink) who he claimed siphoned off too much of his money. He had a new girlfriend now, a local artist, but he admitted that it was often easier without people in his life, except for visits like mine, which he enjoyed. It must have been irritating to face so many questions about happiness, relationships, regrets, questions too private for anyone without such an anomalous lifestyle, but questions that seemed so relevant to Heinz.

We sat that evening, drinking beer, getting onto deeper subjects. My mind was on the refugees. Heinz blamed overpopulation. He thought that the sheer number of people in the world and our reticence to talk about the issue would eventually cause some calamity, a war or plague, 'nature will find a way to correct this'.

'What would you do, Heinz?'

'More birth control. After three children women should be given money to be sterilised; $500. We're crammed in, shoulder against shoulder, fighting for resources. Africans for example, they are *natural* people, they loot and plunder the world. They live for today and don't think of tomorrow. When population explodes like this there will be genocide like in Rwanda, or another AIDS. This is natural!'

I stalled in the face of such anger, unsure where to begin my protest. I pointed out that England has a population density greater than Rwanda, that Britain and Germany had plundered the world more than any African nation, that population had been on the rise while, in broad terms, conflict had been on the wane.

'No! Misery is millions more! They come like raindrops! The NGOs say they are doing their best, but they are not doing their best. Save The Children? No! Don't save the children! There are too many! In Calcutta if the rats get fed they breed and breed and eventually they attack each other. We're just animals.'

Heinz was on his feet now, brow wavy with lines, fists tight.

'Nature will find a balance! Germany should close its borders to the refugees, kick out the north Africans. Let the Syrians stay but send them back later. They cost too much.'

So, Heinz was concerned about the German economy when he'd been an exile for his adult life. He made humans sound self-seeking and cut-throat, unsympathetic, predestined by some biological given, some unchangeable foundation. Foucault thought that asking what defined human nature was the wrong question, that what mattered was how people have used specific concepts of 'human nature' to further their particular views on theology, biology or history. Even if you could define human nature, I'd begun to wonder if our capacity to subvert it mattered more. The very idea of a hardwired nature felt constraining. It reminded me of geographic determinism, the idea that the habits and characteristics of a particular culture are bound by geographic conditions, a view of the world that seemed myopic and fatalistic to me now. If we accept that we are prisoners of our genes, our past or our geography, then we're just spinning through the universe, without agency. We would be resigned, aimless and, more to the point, hopeless.

Humanity cannot thrive without hope; it drives us to better the world. To quote Rebecca Solnit: 'Hope locates itself in the premises that we don't know what will happen and that in the spaciousness of uncertainty is room to act.'

The idea that humanity is destined to fight over resources, that the strongest must survive, that battle is our 'natural state', was the basis of 'Lebensraum', literally 'living space', a concept the Nazis used to justify land grabs and expansion, its principal foreign-policy goal. The global population can't expand forever, we can't endlessly extract and consume either, but Heinz had a zeal for a very specific downsizing that troubled me: curbing some, not others.

'Who are you to define human nature, Heinz?'

'You're a fantasist!' he screamed. 'Six years round the world, it's nothing! What do you know!'

How worldly can you be, I wondered, alone, always moving, never setting down roots? Perhaps living in a community, with all its nuance and diversity, its discordance and compromise, can more powerfully compose a world view (or a just one, anyway) than watching the world as a stranger does, with no stake in anything. This was something I'd begun to resent in my own wandering. The last years had been, for the most part, an exercise in non-participation. I hadn't improved any given slice of the world, but had nosed around, voyeuristically at times, and in my lowest moments I wondered if it had all been a vain search for self-worth and purpose. Perhaps purpose, at least, lay closer to home. Or perhaps I didn't need to agonise over the purpose of life at all, and then maybe I'd die with the simple ideals of Oliver Sacks: 'Above all, I have been a sentient being, a thinking animal, on this beautiful planet, and that in itself has been an enormous privilege and adventure.' Perhaps searching for a purpose is greedy and immoderate. Perhaps it's beside the point, too.

Heinz's rage was fleeting. The next day he waved at me

cheerfully, huffing away on an exercise bike in the hall. He limped off – it looked like arthritis in his hip. There would be no deathbed for Heinz, I thought; he won't go gently. He'll fall from the saddle, no doubt at all.

*

Three days after leaving Heinz, I was in Amsterdam, which meant that it was time to concentrate, and hard. Despite a laudable devotion to cycling infrastructure, it's often hard to understand how cyclists remain alive in Amsterdam. First of all, Dutch cyclists are awful. This might seem surprising at first, given the amount of time they spend on bicycles, but this simply means that all kinds of other activities occur on bicycles too. I counted them: texting, online shopping, dressing and undressing, flirting (I was almost clothes-lined by young lovers, holding hands), reading, eating, smoking, combing and dreading hair. And all on a swarming dynamic of bikes and scooters, while everyone's getting pummelled by gale force wind, tyre-wide tram lines slice the lane at unpredictable intervals, and any tourist in the melee is three days into a mescaline bender.

To cross any road in Amsterdam you need the hypervigilance of a bomb disposal expert. There's a procedure to help, akin to the Green Cross Code: look right, look left, look right again, step forward, get brained by a speeding, lovely, long-legged, upright woman pedalling a cargo bike while on the phone to her boyfriend. The last thing this world will know of you is a soft thud down the line to Lars in Rotterdam. And if you survive, then what? Road rage is unbecoming in Holland: like helmets and fear, not very Dutch.

Having survived Amsterdam, my plan was simple: streak down the coast to Calais where I'd jump on a boat for home, ride for a day and hug my people in London. But as I set off towards The Hague, I began to worry about some desperate anti-climax. Perhaps, having survived Mongolia and Afghanistan, I would die somewhere entirely safe and ordinary, like Belgium. My bicycle frame could snap into seven after summiting a speed bump. I could collide slowly but devastatingly with a six-year-old French girl on a tricycle. I could fall victim to a stampede of sheep on the outskirts of Canterbury. The possibilities were tepid and endless.

A bit like the weather. For the last two weeks, as if to welcome me home, the weather had been all things British, shifting from drizzle, to overcast, to windy drizzle, to Britain's most iconic climatic event: drizzly drizzle. The forecast suggested, though, that I was in for a slapping. A storm named 'Imogen' was throwing a hissy fit and although Britain had suffered the worst of her, she was coming for Rotterdam now. Soon enough, I was battling against gusts of 90 km per hour, which tossed grey herons about like newspaper and launched wind farms into terrific whirling. A man stopped to offer advice, or at least a description.

'It's very windy,' he shouted above the roar.

'I wondered what that noise was.'

'Jaaa. It will be very horrible for you. If you're going to The Hague, it's full headwind.'

'When I drown you in the canal, no one will hear you scream.'

'What?'

'Thank you for your help.'

I crossed from *petite* Belgium into *plus gros* France and it was oddly reassuring that the French were still refusing to recognise any of the syllables I created as belonging to their language. I found the youth hostel in Calais where I'd stayed six years before. Almost empty then, it was now almost full with volunteers working in the Jungle, the sprawling camp of refugees and migrants nearby. After chatting to some volunteers over breakfast, all appalled by the camp's muck and overcrowding, the political neglect of its residents and, relatedly, the dead children washed up on European shores, I decided to pitch in too. Perhaps this was a sign of my desire for absolution: I'd taken from the world; I'd been provided for. For a nanosecond in comparison, perhaps I could give a little back.

I arrived at a warehouse that stored clothes, bedding and other donated items, run by an NGO called Auberge. Hettie, a naturally vocal, high-vised coordinator, was giving the orders.

'Don't go on photographic massacres, guys. These people have been through a lot, respect that. What you're doing is important, even if you feel like a small cog in a big machine. Right, who's good on details, I need five to sort medical kits: one, two, three, four, five. Off you go. Woodcutters! Who likes getting physical? You and you. French speakers please! Go! That is your *domain*!'

With the litter pickers, I jumped into a minibus and we headed to the site, once an industrial dump, passing under a road that carried lorries bound for the UK, their supports scrawled with 'London Calling'. Yesterday it had hailed and the lanes through the camp were mired in mud and blotched with puddles coloured by indeterminate spillages. Shelters were squeezed together in a spread I couldn't see the end of, cheap tents pegged down with shards of broken poles, plastic tarps stretched over wooden frames, pallets used for flooring. A few Portakabins were lined up, excrement piled high onto the toilet seats. We hauled broken tents from the mud.

The lanes of the Jungle were busy with men, tall and Dinka, turbaned and Pashtun, and fringed with Eritrean flags; 'long live Kurdistan' was sprayed onto chipboard. They led to an Eritrean church and a whole Syrian quarter. Refugee, migrant, exile, stateless, displaced, alien, whatever: looking around you couldn't tell the difference, the labels were meaningless, the stories diverse and the segregation a nonsense. Everyone was serving an indeterminate sentence, everyone thwarted, everyone sharing the agony of not-knowing.

The shops were well stocked with fresh fruit, door hinges, batteries, Mars bars. The best of the eating houses were known to be Pakistani and Afghan. The Hamid Karzai Restaurant was a popular choice, and a café run by three Pakistanis from Peshawar called 'The Three Idiots' did a famous masala tea. The White Mountain had an arty black-and-white photo of a shoe covered in mud on the wall. Was the Jungle being gentrified? It was certainly not without a sense of irony. 'The '3-star hotel' was down a muddy track from the '5-star hotel'. There was a 'World Trade Hole'.

Thanks to specialist volunteers the residents of the Jungle could get legal advice. They could perform in a geodesic dome turned theatre, where a guitar player whose hands had been burnt by the Taliban strummed tunes and a professional circus performer from Eritrea did tricks. Anything to help individuals recover identity in a place that could reduce them to 'Afghani in tent'. People could even borrow books from the library; titles such as *Overcoming Depression*, or even the three-inch doorstop *Inorganic Chemistry*.

I walked past more slogans painted onto chipboard:

'I am thinking about the world'

'We live together as brothers, we die together as fools'

'Human after all.'

A sign outside one tent had been stolen from the Eurostar and I wondered if 'Danger of Death' applied here too. I recalled another sign, the first I'd seen in Syria, a wooden board just over the Turkish border, white letters: 'Syria welcomes its dear guests'. The memory had once made me smile, I felt repulsed by it now.

Even purgatory has a schedule. At 6 p.m. on Afghan High Street ('David Cameron Street' had been trialled – it never stuck) the black market began and donated offerings were bartered for. Around the camp there were posters warning of measles and rumours of several cases. Scabies was out of control and almost pointlessly battled. There were injuries from jumping onto trucks, twisted ankles especially, and burns from fires in the camp, which had displaced hundreds. The origin of the fires was as wild with rumour as anything else in the Jungle. Some said it was fascists, others said it was the police or rival factions. Perhaps people simply lit their own tents in a howl of desperation.

I found medical volunteers working out of caravans, axle-deep in muck. Anna, a British palliative care nurse, was cleaning a wound on the forearm of a young, cheerless Syrian boy, without knowing how he'd sustained it, if someone had cut him or if he'd done it to himself.

'A lot of people don't have much in the way of medical problems, you know. Stress causes all sorts of symptoms. They really just want someone to care. We give them socks, some pretzels and sometimes a chat if their English is any good. It's something.'

No doubt some in the Jungle had been raped and beaten into exile, and some form of mental trauma had followed, but there was enough stress here to traumatise in any case: almost everyone had an uncertain future, had split from family and friends, missed home, and was dealing with detention and poverty. I thought about this, as men, and the odd woman, walked past me – each a family fragmented, an international sadness. That the Jungle could be considered merely collateral damage to globalisation and progress felt complacent at best and downright malicious when trudging through the muddy lanes, breathing air that reeked of sewage and burnt plastic.

By the caravans I met Khan, a Pashtun with green eyes and faint spots on his face. He was from Kunduz, the Afghan city east of Mazar that I'd not dared venture to. His father, he told me, had been an academic and was tortured and killed by the Taliban. Khan had 'fought back' and fled.

It was impossible to know how to interpret such stories, as many residents of the Jungle had an idea of which tales to tell in order to gain asylum and I wouldn't find anyone to corroborate. No doubt under similar duress I would lie, shout, rage – you'd be pious to pretend otherwise. Or perhaps such stories were told and retold so often that memories got bent, reality and fiction becoming bolted together. Equally, coming from Kunduz, Khan's tale was credible enough.

He'd have less cause to rejig the next chapter. He'd spent $12,000 to get to Europe, he said, and though his boat had sunk when travelling from Turkey to Greece, everyone on board had been rescued safely. He'd crossed Europe by various means: by foot for twelve hours through Macedonia, by train across Germany. After arriving at the Jungle he'd been detained four times by French police, caught in the act of sneaking onto trucks, spending a total of six months in detention, a waste of time he resented.

Maybe we all travel, in part, to discover the world we want, and what we, collectively, need to do to get there. The Jungle was a sad indictment, and yet the sadness was matched, often quite suddenly, by hope. The volunteers were fired up – you could sense it – sickened by brash tabloid headlines nourishing a moral panic, the 'othering' of mums and dads and grandparents and brothers, of people who like to eat bread and ponder the sky and stretch when they wake and fart and tidy up. Here was a community that on the surface represented a gross condemnation of borders and unequal opportunity, yet it wasn't wholly dystopian. Yes, there was conflict, clannishness and filth, but there was also sharing, trade, friendships, cooperation, banter and laughter. Under the circumstances, you might expect much less. The Jungle was a shitty, confined version of anywhere, and like anywhere, like the world as a whole, like everyone in it, it wobbled between betterment and self-destruction.

Some of the men, women and children I met in the Jungle might have made it to Britain, and if so, some would stay for a while, some would not. Regardless, their impressions would linger for a lifetime. How they were treated as they moved as strangers through the world, through Hungary, where Khan was beaten by police, through Britain's 'hostile environment', through the mucky, stricken slum of the Jungle, this is what would travel next, back with them to Syria and Congo and Afghanistan. Their welcome, or their mistreatment, would change world views, and the very notion of Britain would be trafficked.

This, after all, is what had happened to me during my years as a stranger, in my comparatively cushy, self-inflicted exile. Chieng and Liyan were China to me now. Tariq, who threw me a birthday party in the Syrian desert, he is an ambassador too. His family were some of the few Syrians I've ever known, they shaped the country in my mind, and when, in the years that followed, their homeland flashed burning across my TV screen, I thought of them.

I still do.

Part Five

HOME

Britain … it's basically just a kind of grey, godless wilderness, full of cold pies and broken dreams.

Bill Bailey

18

An Audience of Friends

In a booth on the ferry, there were maps of London for sale, so I bought one, paying in euros, receiving a pound coin in change. Stupefied and over-sensitive, I stared at that pound, lying there in the palm of my hand. More than a coin, it seemed like an emblem of things to come, a little golden promise of a less relentless life, a life with unexciting, everyday things … a bookshelf! Jeans!

The sky was a bright, powder-blue over the English Channel and it was flecked with trembling gulls as we neared the British coast. We docked in Dover and I wheeled my bike onto the ramp, eager to get to home soil, when a man in high-vis stepped into my path.

'Sorry, fella. You gotta wait till this lot get off,' he said, nodding to the trucks. 'Health 'n safety.'

Hoping he'd give me something even more British, I asked him what he meant.

'Well, if I let ya go, see, you'll get run over.'

Not just 'you might get run over' but the promise that I definitely would seemed a more heartening welcome than some measly chalk cliffs could ever have been. It had been six years: I could wait ten minutes, especially if my life was at stake.

I pedalled out of Dover and camped in a field near Kingston, trespassing through an open gate and waking to a frosty morning. A farmer chugged up on his tractor and asked what

I was doing in his field. I explained that I had been sleeping in his field, and now I was eating in his field. Would he like me to piss off? *No, no. Finish your breakfast.*

For old times' sake, I decided to retrace the path I'd cut six years before, and while the A2 was relentlessly hectic, I was cheered up by a pun amid the traffic. A small yellow car rattled past with a large opened-topped receptacle on the roof, and the words: 'Junk and disorderly. All Rubbish Cleared. We do wives, girlfriends, mothers-in-law, husbands, taxmen ...'

Puns! I skipped through the best, or perhaps the worst, I could recall. British high streets are full of them. Fast-food joints came to mind: The Codfather. Tottenham Hot Spuds. New Cod on the Block.

In his famous essay 'England Your England', first published in 1941, Orwell wrote about the cliché, an England 'bound up with solid breakfasts and gloomy Sundays, smoky towns and winding roads, green fields and red pillar-boxes'. He noticed the effect of returning too.

> You have immediately the sensation of breathing a different air ... The beer is bitterer, the coins are heavier, the grass is greener, the advertisements are more blatant. The crowds in the big towns, with their mild knobby faces, their bad teeth and gentle manners, are different from a European crowd. Then the vastness of England swallows you up, and you lose for a while your feeling that the whole nation has a single identifiable character. Are there really such things as nations?

So a pleasant rush of nostalgia can meddle with your memories, but perhaps that's not always a good thing. Writing in the *Sunday Times* on the darker motivations behind Brexit, A. A Gill warned of indulging in half-memories of an old, idyllic, corny Britain. It was a national bad habit, he declared, it was 'snorting a line of that most pernicious and debilitating Little English drug, nostalgia'.

Ahhh nostalgia. It's not what it used to be. During the seventeenth to nineteenth centuries, nostalgia was considered a psychopathological disorder, which was said to cause melancholy, loss of appetite, hallucinations of people and places missed, and occasionally even suicide. The term found fame in the 1688 medical dissertation of Swiss physician Johannes Hofer, and is derived from the Greek *nostos*, or homecoming, and *algos*, or pain, and was thought of then as a frenzied, paranoid sort of longing. During the Thirty Years War, the so-called disease was prevalent in Swiss soldiers, who came down with severe nostalgia when they heard a certain Swiss milking song, 'Khue-Reyen', and so playing it was punishable by death. Copious causes were postulated for nostalgia: a too lenient education, coming from the mountains, unfulfilled ambition, masturbation (that old chestnut), love and an enigmatic 'pathological bone' that doctors were never able to locate.

Treatments were often brutal. When the Russian army suffered an apparent outbreak of nostalgia in 1733 en route to Germany, several soldiers thought to be suffering from it were buried alive to prevent it from spreading. Dubious cures over the years included public ridicule, leeches and purging of the stomach. Clearly, going home would be the obvious medicine, and this was occasionally recommended, although it only seemed to work if the home the sufferers longed for had not changed too much.

Maybe I would be fine then, I thought, as Kent passed by in knotted bus queues, charity shops and slumped, ugly people. But maybe coming home just replaces one nostalgia for another, since already there was some sense of a life receding. I'd miss the minimalism prescribed by the space in my panniers, I'd miss those long, slap-happy days, the luxury of being absorbed in thought, the world unfolding page by page. I'd once read about the Greek philosopher Diogenes of Sinope, who lived in a wine barrel on the streets of Athens. When asked by Alexander the Great what he wanted most in the world, he's said to have replied 'For you to get out of my sunlight!' I felt a little like that when I thought of all the new things I would need in order to slot back into life again. A bookshelf. Jeans.

I detoured through villages when I could, stopping for a pint of Old Dairy ale in a country pub near Kingston called The Black Robin, eavesdropping from a stool at the bar.

'We all know the problem with Clive: he's a lazy sod.'

'Bloody council! Same as always!'

'I ain't drunk 'ere for ages. Not since pie-night.'

Leaving home had been tumultuous, but I felt soothed now, even when that felt like the wrong reaction. White vans skimmed by too fast – the rogues! I stopped in a shop to ask the staff how far it was to Sittingbourne.

'God, couldn't tell ya. Two mile?'

'Nah, 'bout six,' explained someone else.

Customers in the queue all jumped in with estimations, faces contorted in calculation as opinions oscillated from one to eight miles, presumably from people who lived nearby.

'Take bloody days if you wanted to walk it,' said the man who'd suggested it was four miles away. 'About an hour, actually,' I said, and he looked at me like I'd suggested a free dive to Hades.

Sittingbourne was a vital stop in my journey down memory lane because it was home to Tommy and Roger, the kindly strangers who'd taken me in when my journey was 36 hours old, the first of many. Tommy opened the door – 'Stephen!' – and I was quickly hustled inside to the cake they had baked in my honour. I must have been more dilapidated than they remembered me, and I probably smelt of mildew and feet, but a bed was made up and I crashed out smiling, covered in crumbs and the grime of the A2.

Subconsciously, in my dewy-eyed excitement for home, I'd been editing Britain, giving it a nice polish, rubbing out the flaws and frustrations. Like bike lanes. The ones in Kent were woeful things, landscaped by tussocks, laced with tree roots, muddled by random bits of curb. They disappeared under wheelie bins and parked cars, or for no apparent reason at all, other than, perhaps, the council running out of paint. It reminded me that home is home, warts 'n all, and that it's the ugly beast of patriotism that makes nations bespoke, that grants permission to take the nice bits, Yorkshire pudding, say, or universal healthcare, and leave the rest – historical imperialism, race riots, Piers Morgan.

It is a hackneyed notion that travel lends itself to self-discovery and I wondered now if it applied to me at all. A solo traveller can only scrutinise themselves, and perhaps scrutinising your relationships with others is a safer mode of self-discovery. 'Man only knows himself in so far as he knows the world', said Goethe. Sometimes, we know things best when we take a slanted view – could this work for my impression of home too? Do you best comprehend home by looking elsewhere, by familiarising yourself with other places? The Polish writer Ryszard Kapuscinski thought so. 'For your culture will best reveal its depth, value and sense only when you find its mirror reflection in other cultures, as they shed the best and most penetrating light on your own.'

This happy idea came to mind as I followed the Thames estuary, as metropolitan London unfurled in halal butchers, a Chinese medicine shop, an Indian curry house, a bingo hall, countless pubs and a fish and chip shop, serving our national dish, an innovation of Jewish refugees from Portugal. There were faces with hints of all the countries I'd been to and many I hadn't, countries forever twined with British history, like it or lump it, part of a great ongoing cultural diffusion, abetted today by travel, migration and the internet when once it fell to explorers, merchants, soldiers and slaves. I wondered about how the great tapestry of British culture came to be, what had been weaved in and left out, what had been embraced and celebrated, diluted and simplified, and what had been blamed for a national identity crisis. Another pun came to mind, rollickingly British: *Jamaican me hungry*. Pretty good.

From Shooters Hill, London spread out, tangled streets blotted with shifting traffic. I dropped down towards Blackheath, where hipsters on fixies with D-locks looped through the backs of their jeans flashed by my shoulder. In Bermondsey an empty can of Special Brew stood aloof on a railing, like an artwork. I stopped at the Cray Side Café for a second breakfast, a full English with bubble and squeak. The TV in the corner showed David Cameron engaged in talks with the EU. An interview began with a single mother turned kleptomaniac. Near Greenwich there was a commotion outside a chemist, a thief sprinting, bits tumbling from his bag, two old ladies peering down at syringes, scattered. And fuck, it still felt great to be home.

From the northern end of Westminster Bridge, I could just make out a gathering outside St Thomas' Hospital, friends and family waiting for me to return at the appointed time. This was it: full circle. I stopped, took a minute, curious about how the next moments might play out. I'd thought about it plenty of times. In one rather self-aggrandising reverie, I got lost in the kind of hooting crowd that welcomes home a hero in a cheesy American film. It might not be so glorious. I might cry. I might turn into a blubbering embarrassment who has to be led quietly away. I might not come home at all. I could spook, hang a left, ride down York Road, set a course for Chad.

But when the moment came, I knew Chad would have to wait. I felt tired. Not bored and jaded, not wearied by the world, but tired of my long-winded, insistent motion through it. Tired of waking befuddled in my tent and mentally pursuing my place on the planet. Tired of the newness of my friends, the oldness of my socks and the staleness of my bread. I was raw-boned and sapped, to say nothing of my buttocks.

I crossed the bridge and rolled around the corner towards St Thomas' Hospital. Friends and family had gathered around a finish line. Someone handed me a bottle of Prosecco and I hugged my mum. Nobody hooted, but in the first of a relentless stream of comparisons, I realised that I hadn't been the subject of so much attention since rural Azerbaijan.

A couple of boozy days later, I began to cycle back to my mum's house in Oxford, sticking at first to unpaved towpaths, the route suggested when I pressed the bicycle button on Google Maps. I hadn't been so mired in mud since the Pamirs.

Towpaths turned to roads, and, after Thame I flew through a run of three villages, each a mile apart, and which must have selected their names in the hope that no outsider would ever come to visit. Shabbington, Ickford and then Worminghall. I looked to see if there was a Lake Sputumere, but couldn't find it. To kill the final hour, I tried to come up with taglines for their signposts.

'Shabbington. Please remove your coat tails.'

'Ickford, twinned with Ewford.'

'Worminghall. Drive slowly. Invertebrates at play.'

And then I arrived into Oxford and that was that, a stirring, testing and selfish chapter of my life was at an end.

I had crossed 102 international borders and a similar number of cranky immigration officials. I had filled 23 journals cover to cover with weary scribblings. I had worn 26 tyres, sixteen cables, fourteen chains, twelve sets of pedals and five Rohloff hubs to scrap. I had camped for free by roadsides for over a thousand nights, and of all the statistics I dutifully collected, this is my favourite because it reminds me of the capacity for freedom in a world more dependable than I'd ever imagined. My final tally was 53,568 miles* of bicycle travel, but none were longer, or more daunting, than the first – those half-lit miles between a South London boozer and a moth-eaten guesthouse in Bexleyheath.

It is typical, having completed such an expedition, to put the distance into perspective and make it sound striking, so I will. I'd pedalled a distance equivalent to more than twice around the equator, or 61 times Land's End to John O' Groats or 24 times the Tour de France or 23,808 laps of the Coventry ring road or almost a quarter of the way to the moon or 9,743 Mount Everests, if you'd somehow cloned the mountain, laid it end to end in space and pedalled through its rocky centre in a burrowing bike-human space suit. Choose whichever analogy sounds most remarkable. I like the Everest one.

And yet most of those miles count for so little. They are just numbers. They reveal nothing more than a steady physicality and they hardly speak of progress. I'd taken a serious detour around the world, making ups and downs and arcs and hairpins, just a long scribble on my map, but as lovely as a spiral, which goes out of its way while moving forward too.

I was proud that I'd stuck it out for six years, and proud too of the depth of faith this required, but I was humbled by how it was made possible. Countless strangers had given up beds for me, in almost every nation, and I'm willing to believe it would have happened in Monaco and Bosnia too if I'd stayed long enough (well okay, maybe not Monaco). And I'd slept in

* It would have been 53,567 if I hadn't got lost near Greenwich on the way back into London.

a hotchpotch of other spaces, in schools and police stations, in hospitals, churches, mosques, temples, monasteries, fire stations and army barracks, and I'd been thanked for accepting some of these invitations. In Egypt, I'd shared the fusty air and mosquitoes of a barn with a snortsome buffalo. I probably wouldn't ever do that again, and, for reasons I couldn't fully explain, that felt crushingly sad.

I thought back to my old, bumptious mission statement: 'cycling six continents'. My hopes and goals had shifted on the road, and I'd come home with better reasons to leave. Travel is always a wildly biased enterprise – not a bad thing, just a fact – our view will always be blurred by prevailing myths and clichés, tinted by our experiences, and our place in the world. Along with my tent and clothes and toolkit, I'd been lugging some hefty cultural baggage around with me, and not even the humble business of cycling could change that. But I'd been seeing as a doctor too, and, finally home and in reflective mood, I felt grateful for this particular lens most of all.

19
Rehab

What happens when you treat wanderlust with travel? Can you purge it from your system? Well, not always. For a few rudderless souls, for the likes of Heinz, travel seems to exacerbate it. For them, wandering is like muscle memory – a movement, repeated over time, which at last comes naturally. Happily, then, my own dromomania was finally in remission. What's more, I was not the monosyllabic ghost that I'd half-joked would be the product of such a slow and relentless journey, most of it alone. Of course, this was my immediate sense of things, and it was just the honeymoon. After a few fallow weeks in Oxford, a faint sense of panic set in.

I began to feel as though I'd ceded something important, something beyond a spirit of pursuit. I felt discrete, less rooted in the world. For the last few years, I'd often been asked about the future, and usually, it was a variation on: *How will you cope once you get home?* Sometimes, I sensed a note of concern here, or an implication that I was a condemned man. Was I? Could I be institutionalised to The Road? Perhaps I'd cope by going away again. Perhaps I'd set off to work on my bike one morning, a short commute through London, never to arrive. Instead, my mum would receive a phone call: the police had found a panicked man meeting my description 260 miles away, trying to construct a lean-to in a ditch using his shirt and tie.

What *do* you do after a six-year bike ride? I couldn't dodge the question forever. Aside from writing this book, none of the obvious options held much appeal. Briefly, I considered merging my adventuring and medical skills into the role of expedition doctor, and nothing sounded less glorious than treating Gap Yahs for blisters, cold sores and mild anxiety. I could be a speaker, coaxing others towards more deliberate and meaningful lives. This was a well-trodden path for returning, ahem, 'adventurers', though for a mildly sad, single, debt-ridden 35-year-old, more calf muscle than man, who lived with his mother – and blogged, for fuck's sake – I just wasn't convinced that I should be the one giving pointers. And anyway, I found most of these motivational types weird – they're the sky-shooting extremists of optimism, excitable man-kids, mostly, cult-leadery and sickeningly smug.

I did not to rush to reintegrate. I kept clothes in my panniers for a few weeks despite having a wardrobe, and occasionally used a sandal to hold my cup of tea in position, for old times' sake. I still ate frantically too, but now, with my mum and her partner George watching on, I'd half hear things as I did so.

'When do you think he'll stop?'

'Dunno. Give him more, let's see what happens.'

Apart from the frame of my bike, only a single, threadbare blue T-shirt had survived the whole journey, and it was like a comfort blanket to me now. One morning, I turned my shoe upside down and shook it through sheer force of habit. Scorpions, though, aren't such a problem in the home counties. Abdul from Sudan said hello on Facebook Messenger and we got into a circuitous conversation. 'How are you?' 'I'm good and you?' 'I am fine, and you?'

I turned wistfulness into a prolonged and serious occupation, arranging a few cowries and cone shells from Costa Rica on my window sill, a pine cone from Queensland, an ink etching of a hummingbird given to me by Edit, a Hungarian artist who I'd fallen for in Budapest. We'd spent a week blinking into each other's eyes, her hands running through the last of my hair, listening to what would be a toxic amount of Bobby Womack under other circumstances. It occurred to me that I'd been a time traveller of sorts – for years as a doctor in London I'd dreamt endlessly of the future. On the road, I'd landed hard in the present. Home again, I lolled helplessly in the past.

Oxford had changed, superficially at least. The chippy flogging sodden chunks of cod in newspaper had gone, replaced by a bistro joint. The newsagents had been cleared out too, and Costa had moved in. On the benches where weaselly, hooded boys once sat cramming skunk into king-size Rizla, a pleasant Macedonian man sold seafood paella out of a huge wok. Life moves on.

One morning it hailed out of a blue-grey sky. Ice tinkled against the double glazing and skittered down the roof. This would have virtually no impact on my day, and I lingered glumly on how unaffected I was now by the whims of the outside world. The sky could have been quartered by fork lightning and the most that would have happened would have been me going 'oooooh' at the window, never worrying that I might be electrified to a lonely death, like the good ol' days. I didn't – couldn't – find anything even remotely terrifying now. My life would remain nice and comfortable, unless Pret A Manger ran out of crisped kale.

I found some solace in my friends, and, to my great relief, they were still the friends I remembered. No one was wanted by Interpol or the FBI. Nobody had developed a counter-cultural fondness for, let's say, yodelling, cockfighting or burlesque. The closest any of them had got to transformation had been, in one case, getting drunk enough to get dice tattooed on his arse.

My mum, meanwhile, was smitten with an adventurer who wasn't me. Levison Wood was doing venturesome stuff on

Channel 4 at the time, writing books about his escapades in Africa and the Himalayas, giving public talks with aplomb.

'He's such an adventurous guy,' my mum said.

'Mum,' I began, steadily. 'I've been cycling around the world for the last six years …'

'I know, I know,' she said, before lapsing into a reverie.

'But Lev is so handsome, isn't he?'

'Yeah,' I said. 'And he sounds like a golf club.'

She followed him on Twitter, and I wondered when she was going to follow me.

'Oh, are you on Twitter? I didn't know,' she said when I reminded her. I sent her my handle, but she was lost to Lev's feed, embarking on a festival of 'likes'. One day, she insisted we see him give a talk about his adventures in Oxford. She told the usher that she couldn't sit on the balcony because she had vertigo – not something that she'd ever suffered from before. We were given seats at the front, and my mum looked completely satisfied, closer to Lev.

I was able to spend Mother's Day with my mum, for the first time in seven years. Unfortunately, this coincided with the day she took me shopping for clothes. I wasn't able to get back to work in the hospital just yet, and I was broke, so she paid for all my stuff. I clumped along behind her in Sports Direct as she held up bargain-basement jeans and said 'try these on, Stephen'. I was the only 35-year-old man on the premises in this position, though I briefly locked eyes with a twelve-year-old in a similar one, trailing his own mum near the checkout. We swapped a short look that said: *Pfff. Mothers.* As my mum paid, I could see the checkout staff thinking: what's wrong with *that* guy?

What was wrong was straightforward enough: there was a cost to what I had done. Worse: I had no right to be irritated or self-pitying about this – it was my own mess and I was pissed off with myself for neglecting to consider or prepare for it. I'd left home in my twenties, unwilling to accept that opening some doors closes others. You could call it being entitled I suppose, or credulous, at the very least. I'd believed the advertisers, social media blowhards and all the other glib correspondents of modern life when they preached that you can have whatever you want, free of charge. Obviously, that's bunk. You can't have the light without the shade. I hadn't fully mapped out this part of my journey: the cohabiting with my mum while rapidly going bald. My debit cards and credits cards were maxed out, the Student Loans Company were sending threatening letters, as if they might repossess my shoes, and I was now living on cut-price baked beans, the kind that taste like salty cork. It's not that I hadn't expected *some* fallout, just not loan sharks and scurvy.

There is, however, a wonderful cure for the pain of a long journey coming to an end. And that's to journey some more. Three weeks after coming home, I zipped off to Singapore.

The invitation arrived by phone from an affable corporate man with the voice of a radio DJ who'd read an article I'd written for an adventure travel magazine. I was to serve as an inspirational speaker, one of those smug chumps. They were offering a fee that persuaded me to be as smug as required. There was only one drawback, really: I would fly to Singapore, which meant covering in twelve hours what had taken two years to cycle.

It was a small world from the window of the Boeing 747. I turned to the digital map in front of me and watched as the pixelated plane cleared the Uzbek desert and its millions of camel spiders in what felt like minutes. In the arrivals area of Singapore airport there was a man holding a sign with my name on it. So I was one of *those* people now, people who have their names on signs in airports! I felt a fierce bond with this stranger holding my identity in his hands. He didn't want to chat though, and I was quickly ushered outside. My car was a polished, black BMW. My driver opened the door. I flashed a smile at a hostess from Singapore Airlines, ducked inside and supped mineral water as the ugliest brand of smooth jazz ever composed played on the stereo. This was the big time, baby.

Singapore passed by the tinted windows, still reaching, glitzy, futuristic. It was another chance to marvel at the unlikeliness of it all – a city that grew from nothing, with no natural resources to speak of. Yet again, it had been reported, by the BBC, that it was the most expensive city in the world to live in, so it was just as well that the company was to pay for all my expenses, including the five-star Swissôtel The Stamford (once, in the 1980s, the tallest hotel in the world).

As I stepped out of the Beemer, a preened lady appeared immediately. 'Dr Fabes – let me show you to your room.' I sauntered through the hotel lobby, where abstract impressionism adorned the walls, and stepped into a lift beside two Russian oligarchs with huge biceps. I was on the 56th floor and we ascended so fast that my ears popped.

When you've spent most of the last six years in a small, dank tent, this is what happens when you arrive into your own five-star hotel suite … First, you starfish on the bed, like you're an actor in a TV advert for a hotel chain. Then, you steal a few bedroom items for the sheer fuck of it – stationery and soaps, mirrors, whatever you can fit into your baggage. Then you undress and stand naked in front of the twinkling expanse of Singapore, with your arms outstretched and your bits exposed to the city. Then you use the first name of the bellboy when thanking them, reading it off their name tag, like a wanker. Then you push buttons for a while, turning on every electrical item simultaneously and turning the air con down to minus 80. Then a man knocks on your door, asking if you want a local or international newspaper delivered to your room each day. You demand both and snort at him for asking.

I stood naked on the bed and twirled a towel above my head like a cowboy's lasso; it was as big as a curtain and as white as a story-book cloud. I eyeballed the mini-bar menu and did some quick arithmetic. 'Eight quid for a tiny can of beer!' I yelled in delight. The menu said that 'Guests may enjoy their favourite aperitif' – I didn't even have a favourite aperitif!

There was a balcony but unfortunately the door to reach it was locked. Later, I learnt that this was because melancholy billionaires occasionally threw themselves off instead of taking the elevator. One had landed quite recently on the pavement below, near McDonald's, which presumably made someone's happy meal a memorably unhappy one.

And then a curious thing happened as I thought about free-falling billionaires. The Stamford began to feel less luxurious. *Huh*, I thought, as I tossed mini bottles of vermouth at the widescreen as if playing posh darts. I thought back to the mornings I'd spent not so far from here in the Malaysian highlands, the sun a fiery circle on the horizon, bringing colour to the world, all mine to admire from my sleeping bag. Somewhere, perhaps 100 or 200 or 500 km north of here, in the palm plantations, there would be a cycle tourer scouting for a clear patch of ground to pitch their tent, and I envied the little pauper that.

At least until the all-you-can-obliterate buffet. A little voice in my head said 'play it cool', but that took half a second, and by that time I'd played it extremely uncool by turning my plate into a massif of incompatible foodstuffs. Sushi got a dressing of beef bouillon. Smoked salmon sat in a lake of – what was that, anyway? Thousand Island?

There was an antidote on offer for the gluttonous: the hotel had a gym, and I owed it to myself, after weeks of inactivity, to go to it. It was awful that I had a choice to exercise now; my brassy new life came burdened with guilt. The gym was like all gyms, full of vain, beautiful people, whose vanity was the cause and effect of their visits to the gym. First, I yanked away on a rowing machine, and then, reverently, I moved onto an exercise bike. It was a fall from grace – how could it not be? – from round-the-world bike ride to spinning. And even more spectacularly, given the pre-set workouts. One of them was – and I'm not making this up – 'cycling around the world'. I set off on this virtual tour of the planet as fake mountains rose and fell on the screen, my heart slow-clapping along. Bored, I switched to manual mode, set it to maximum incline and effort, and got to work, sweating all over the machine and floor, my armpits raining like a Malaysian typhoon.

Travel, eating and exercise were entrenched habits, not easy to break. A session in the gym had to be an all-or-nothing affair now, with at least some risk of emergency cardiac surgery, or I might as well have stayed throwing vermouth at the widescreen. I needed that buzzy hallucinogenic adventure that you achieve only through exertion. It's like briefly joining another world, where voices shimmer and colours sing. The exercise bike had a digital display of my heart rate and I wondered if my final fleeting vision in this life would be of the sequential figures 202, 705, ?#, ERROR, 802, 0.

But nothing so dramatic happened, nothing but my reflection growing sweatier in the copious mirrors all gyms have so that the vain can gorge on themselves. There were no landslides, downpours or ferocious dog chases. It just wasn't the same. Surely, I didn't miss the traffic? I did, a little, I did.

*

Home from Singapore, I moved to London, but I was lonelier in the big city than I'd been during those six years of living estranged. I could hop on stage and spiel away about my ride at adventure-themed events, but I was anxious around people, especially in groups, more so than I'd ever been before. We think of our personalities as solid qualities, but they are fluid, and the more introverted you behave, the more you become so. It would take time to readjust.

My mind had meandered unbidden for years, so initially I found it tough to concentrate on anything. But, determined that rehab was a battle I would win, I made variety my new goal, as much of it as I could squash into this new life, terrified that any less would feel like sinking. Newly passionate about my work, I took jobs at two hospitals across three different specialties, mixing up my shifts so that they fell on different days and at different hours. I consoled myself that I was free to book a flight to Lesotho at three in the morning if the mood took me. I began to write, heeding Laurie Lee's account of autobiography as 'an attempt to hoard life's sensations'. People were dating using apps on their phones now: I joined four. To remedy the comedown of my exercise addiction, I joined five different running clubs too. I ran 70 miles a week, travelling to fells and trails and foreign cities to race, until my nipples bled, my toenails turned black and dropped off, and I could run half marathons in under 75 minutes. After a couple of years, I missed my target in a marathon after suffering calf cramp at 22 miles, and, though I still crossed the line in 2:41, I beat myself up about it for weeks.

I ran to feed the dopamine beast, I ran to solve problems, but, perhaps mostly, I ran to escape. It's a classic tactic. Read the blurb of any running memoir and, chances are, the theme will be 'Running saved me from ...' Insert heroin, bipolar disorder, my intense/life-sapping job, my stale relationship, unrequited love, trauma, obesity, myself, my past. You don't have to be a psychoanalyst to see why running in particular took hold. Running saved me from the responsibility of a new life, new plans. It was a sponge for obsessiveness, a release, and a physical manifestation of a restlessness inside. Loving it came later.

One day I turned to Crazy Max, who was sheltered under a tarp, still muddied from the towpaths. He squeaked rambunctiously when I wheeled him onto the road. The cycle tourers of old had similar pangs of nostalgia. Fred Birchmore wrote this of his bike Bucephalus after he returned home: 'The only thing that seemed real to me was old Bucephalus, a battered and bruised wreck of his former glory, but still standing staunchly beside the door as if waiting for the command to continue our march.'

Crazy Max and I marched off for the Cherwell river valley. It was a Sunday, mid-May, and the warmest day of the year so far, the kind that inspires Britain to revoke its policy of wearied grouchiness. Pub gardens were packed and sunbathers affirmed that life was a cinch.

I rode on lanes beside rapeseed as wood pigeons cooed. The air smelt of hot summer grass. There was a glimpse of sunlit corn and then back into a forest of nettles, which whipped at my knees until I hoisted my feet onto the top tube to avoid a sting. Near Woodstock I joined a bridleway full of green light, and it wormed between neglected fields, the cow parsley high and intruding, like plates of canopies. Cabbage whites danced about humpy canal bridges, which I took at speed. As the spire of St Mary's in Kidlington grew up over a meadow, the shadow cyclist appeared again, rushing through the long grass and I swear I heard him cheer. I could keep going, find another trail. I could gorge on fruit, doze under a tree. I could wake up and watch the clouds roam and billow. But knowing that I could, that the world fizzed with such possibilities, seemed enough for now. I left my daydreams in the field and headed for home.

20

On Us

I arrived home in early 2016, that trembling year. A turning point for me, a tipping point for the world. It was the year an armed man walked into a gay nightclub in Orlando and killed or wounded over a hundred people in the deadliest terrorist attack in the US since 9/11. The year a 19-tonne cargo truck driven by another estranged extremist ploughed through hundreds of people in Nice, killing 86. The year a British Member of Parliament, Jo Cox, was shot dead by a far-right terrorist with a sawn-off shotgun outside a rural library.

It was the year of the Brexit referendum and the election of a celebrity tycoon as US president. Nativists were getting the airtime they craved; the online, attention-economy playing its part. Billboards appeared across London with a warning, 'Breaking Point', alongside a photo of queuing Asian refugees crossing the Croatia–Slovenia border, aimed at stoking fear into the hearts of Britain.

Within eighteen months of coming home, two terrorist attacks had hit London, and both involved my NHS Trust in particular. One was a vehicle assault with mass casualties on Westminster Bridge, one hundred metres or so from St Thomas' Hospital, on a day I was working, ten minutes after I'd been walking over the bridge myself. Less than three months later, on London Bridge and close to Guy's Hospital, a van swerved into pedestrians. Three men got out wielding knives and began a killing spree. Kirsty Boden, a 28-year-old theatre nurse, rushed to administer first aid to a victim. She was killed too. Stabbed in the chest with a 12-inch blade.

At the time, I sensed Britain was changing, insistently, swiftly (a slightly strange idea now, given the more seismic shifts of 2020) – or perhaps this is how it always seems, when you're back home at last, after a long spell away. But I sensed a mood of division now too, and it was harder to write this off as a trick of perspective. It was everywhere I looked: online and in the papers, with views hurled from every angle – emotional, entrenched, extreme. Identities were being touted as vital, cast-iron things. Discrete opinions were out, demands to pick a side were in. The language echoed that of conflict – you're an enemy or an ally. Cycling around the world had brought our commonalities into focus like nothing else could and yet now, at home of all places, there was talk – or insinuations, at least – of deep-rooted differences between human beings.

Were there? To borrow that slippery catchphrase of the Khao San Road, I preferred to think of us – all of us – as *same same but different,* or to put it another way: as similar most of all and diversely different. Perhaps this is because I'd spent a great deal of time contemplating healthcare as I travelled, and this had stressed our similarities and trivialised our distinctions, dealing, as it must, with the raw machinery of our humanness, our blood and bones. I had been privy to a great deal of everyday life from the saddle of my bike too; a reminder, if any was needed, that the fundamentals of all our lives are the same. With such things in mind, I began to wonder if paying more attention to all that we have in common might pave the way to a brighter future. Perhaps this is not an obvious conclusion to draw. After all, the traveller hopes to marvel at how different people, and places, can be. Travel is obsessed with *otherness*. Thing is, I'd seen a lot of *othering* too.

It was never far from view: the hounding of people for reasons as immaterial as they were various, to do with race or mental turmoil, and for such trifles as bacteria, or lines on a map. And from mental health rehab in Mumbai to a decrepit sanatorium in the hills of Georgia, I'd seen othering knitted into systems and societies. But I suspect that it is born, for the most part, of snap judgements. I don't mean to be too hard on us. We're all biased, and nobody is immune. We all need preconceptions of one type or another, they help us pilot a path through the world.

Perhaps the tendency to judge one another on superficial differences, on fragments of identity, resides deep in our genes, but then perhaps there's a sense of fairness in our double helix too. I saw spectacles of unity and acceptance as well, on the road and after my return. Even in cantankerous old Britain, there are a handful of places where we can all come together. A delayed train – when passengers let the usual rules of engagement drop and can have a good moan to each other. Free festivals, when the community marshals in all its mish-mashed glory – there should be more of them. And the emergency department waiting room: where we all sit together, whether the drugs aren't getting you through the nausea of chemotherapy, or your dad's got chest pain again, or you've fallen off a bus after one too many Red Stripe. In the emergency department, we'll help you – money and status aside – and until then you'll sit together, you'll notice each other, you might even strike up conversation. You'll appreciate that Britain can be odd and various and funny and angry and desperately sad. Nothing in this world can help forge connections between people like pain can, or almost nothing. Perhaps only hope can do better, but we'll get to that.

Cycling around the world wasn't exploration, at least not in the traditional sense – we've tramped over most of the world's mountains, deserts and tundra; over the centuries, the physical planet has largely been 'explored'. But let's broaden that out. There are other ways to explore. In a surgical sense for instance, an exploration is an investigative operation. Exposing connections ('the inescapable network of mutuality', as Dr King put it) had been a kind of adventure too. On the way, I'd been awed by wild landscapes, and by a common generosity, but most of all, by the tremendous complexity of the world. I'd been similarly awed once before, as a lowly medical student, exploring the complexity of the human body, our inner world.

If you believe we heal the body through medicine, in the broadest sense of that term, and the world with politics and diplomacy, you might notice that both medicine and politics are becoming increasingly complex – medicine: with each new-fangled drug and technology on offer, with patients living longer to suffer more disease; and politics: as we become ever more knowledgeable, populous and interdependent. But we over-simplify: it is a global and old and very human bad habit. When we obsess over identity and type, over categories and diagnoses, we're in denial of our glorious complexity. When we care so deeply for lines on maps, inevitably, something will sweep over them without noticing our obsession: a volcanic ash cloud, a contagious disease, weather, ideology, misinformation, and we'll feel foolish for trusting so much in convenient fictions, for isolationism is not simply doomed, it is a fallacy.

Perhaps it was cycling's fault – so slow and meditative, so surreptitious – but I arrived home with a heavy sense that we wish away the details, the fine anatomy, of the world. It's the homogeneous portrayal of the African continent in Britain, the nationalistic drive in China for a single identity at any cost, the surface-level care we lend the mentally ill. It's the debasing headlines of Aussie or Kenyan or British tabloids, it's the treatment of vulnerable and persecuted minorities in Myanmar and China and Calais. Call it prejudice or bias if you like, but what's bias if not a symptom of our desire for a simple, consistent world, where people do expected things?

Today, there is a 36-year difference in life expectancy between nations. Great rifts in health outcomes exist within nations too. Healthcare workers are not evenly spread – low-income countries have ten times fewer doctors per head – while healthcare costs drive the sick further into poverty. And all the while, dangerous overdoses of wealth and power are deemed necessary, or are explained away as the collateral damage of progress.

As I fiddle with the final words of this manuscript, Covid-19 is declared a pandemic by the World Health Organisation (and, tellingly, a 'foreign virus' by the president of the United States) and I am writing this from the uncomfortably uncertain place of not knowing how things will pan out – how deep the trauma will run, how severe the reverberations: social, economic, lives lost. I only know that healthcare systems, like immune systems, will be overwhelmed – and it is too early to say if the pandemic will heal society's wounds or open them, but I suspect that this is a false dichotomy, and that it will, in different ways, do both. But as the virus tests our unity and resolve, I am reminded that societies are still not yet fully accepting of the fact that disease plays out on a social stage, has social drivers and social implications. Underfunded public health, under-staffed hospitals, inaccessible healthcare: these things can have tragic consequences, but especially in times of crisis. The most vulnerable will suffer most, in their health and means and capacity to act. Some will be made scapegoats. And yet a virus doesn't see us as we see ourselves, as different. To the virus, we are as one. And yet we will not all suffer the same disease, the same physical and financial hardships, the same grief, the same fate.

So the world goes into lockdown and international travel is

curtailed, yet it has always been so for the most marginalised people of the world, who have restrictions, whether politically-fuelled or economically-set, whether racist or otherwise, on their movements around the earth. And in the coming weeks and months, intensivists may have to make desperate decisions as to which patients get ventilators, whereas inequality, greed and social bias conspire to decide such things for many people, who don't have access to universal healthcare, let alone intensive care units and flashy machines to enhance their chance of survival. I think again of the utility of borders as we set about cornering the outbreak (too late, in most cases), but also of their tendency to obscure the vital issue of our interdependence. The spreading infection is yet another reminder – and I've been granted many – that health does not sit aloof, impervious to the rest of our world; it depends on myriad, interrelating forces, from wealth and poverty, to the health of our planet, to who we call friends and who we exclude. Perhaps the real fight is not *against* a virus, but *for* global solidarity, for we can achieve so much together when we weaponise science and information, when we spread empathy and respect, when we balance caution with hope.

But sadly, whatever the challenge, be it pandemic, war, mental health or mass migration, we close our eyes to complexity, and now I am pondering why. Perhaps it's because with complexity, comes uncertainty, another unpleasant and frightening fact of life, and one that, 'in the presence of vivid hopes and fears, is painful, but must be endured if we wish to live without the support of comforting fairy tales'. Bertrand Russell framed the challenge.

A world without fairy tales is unpredictable, like people. This makes practising medicine (and travelling, and love, and scrambling around the outdoors, and almost anything worthwhile) exhilarating and excruciating in equal measure. As a doctor, I've been surprised so often – patients have died when they shouldn't and survived when they've been ordained to die. Theoretical medicine and real-world medicine are not the same thing – you learn this early and it's important. Patients rarely walk into emergency departments as textbook cases, using the language that you've been taught to expect.

'How would you describe the pain?' I'd asked an elderly man in my final year as a medical student, hoping he'd conform to a list of possibilities sanctioned by my Clinical Studies lecturer. He could run with 'burning', 'crushing' or 'sharp', he would kindly grade it out of ten and explain where it 'radiated', nudging me towards a diagnosis.

'I'll tell you what it's like, doc. It's exactly like ... you know when someone removes all the marrow from your arm bones and fills them with ice crystals?'

Medical textbooks gloss over the vagaries. The pig-headed decisionists who govern the world today do so too. Perhaps, again, there is something instinctive at work here – we're certainly psychologically prone to the buzz of anger and dread; we're all suckers for fearmongering and fairy tales. So maybe we'll always crave over-confidence in our leaders and on we'll go, voting for whoever bears the snake oil of certainty.

But maybe not. Thinking about today's politicians, I see parallels to the doctors of the past, who were once expected to be paternal, cocksure and autocratic in a similar way. To a large degree this has changed, and continues to do so. They used to say – and some still do – that a surgeon can be wrong but never unsure. This strikes me as a bone-chilling lack of humility. The best doctors I know are closer to guides than dictators. Decisiveness is necessary of course – nobody wants a faint-hearted doctor – but the trick is to find the sweet spot: not too certain, not dithering. In the early phase of an emergency – be it cardiac arrest, or pandemic – it pays to hustle, and to act with conviction. But otherwise, listening, remaining open to other views and ideas, keeping a level head, measured and collaborative thinking – surely this is how you best care for your patient, or for the citizens you represent.

When we model politicians, we might keep in mind how patients can be harmed, or neglected, by their clinicians. Patients suffer when doctors overlook or misinterpret clinical signs, when they're lured in other directions by a biased view of things, when they're adversarial, or when ego kicks in and they refuse to seek help or advice. Being a good doctor, they say, boils down to just four things: shut up, listen, know something, care. This could work, I suppose, not just for doctors, but for any of us.

So that's it, my two cents' worth. Cycling around the world was, for me, an object lesson in the importance of accepting intricacy, and of holding our biases lightly. I see such value, now, in connections, empathy and hope. And I am hopeful. When I ponder what I've learnt about the human body, it's not its flaws and vulnerability I find most striking, but its resilience and adaptability. I've been similarly inspired by these traits in the wonky world of today. John Foster Fraser and Thomas Stevens pedalled around a very different planet at the end of the nineteenth century – they found people piebald with leprosy in China, malnourished children everywhere, and numerous diseases devastating whole communities, the causes of which were not understood, let alone the means to treat or prevent them. It's easy to wonder at our progress since then, and today there are calls, indeed whole books, asking us to be more optimistic about the future. But why should the best we've done so far set the standard? Just as hand-wringers can hinder progress, so do the contented. We would condemn our world to stagnation if we were all satisfied with how things are. In many emergency departments, including my own, nobody uses the 'q' word (quiet) when it appears so: more than a hex, it's complacent and self-congratulatory. There's always more work to do.

In his commencement address to UCLA medical school, the writer and physician Atul Gawande asserted that 'curiosity is the beginning of empathy' – this seems to me one of the most important truisms of medicine. If you listen, a patient may move you to see a little of their world, and your view of the world at large may shift a little, or expand a little, too.

I'm back working in the emergency department again now. In the waiting room, I call out the name of a patient. For a second, nobody answers, then an old woman with her back to me prepares to stand. It takes her a moment. Another woman, maybe her daughter, picks up her bag and guides her with a hand on the question mark bow of her spine. In a cubicle, we each take a seat. I don't know yet what's drawn her to hospital, or if we'll be able to help her or not. I don't know how her years and experiences and genes and place in the world have composed her, but I can't wait to learn more. I pull my stool in closer and ask her to do something for me.

'Tell me the story.'

Selected Bibliography

Part One: London to Cape Town

Arnold, David. *Imperial Medicine and Indigenous Societies.* Manchester University Press, 1988
Austin, Henry Herbert. *Among Swamps and Giants in Equatorial Africa: An Account of Surveys and Adventures in the Southern Sudan and British East Africa.* Forgotten Books, 2018
The Bicyclist's Pocket Book And Diary. London, 1879
Brown, Monty. *Where Giants Trod: The Saga of Kenya's Desert Lake.* Quiller Press, 1992
Cahill, Tim. *Hold the Enlightenment.* Vintage, 2003
Camus, Albert. *Lyrical and Critical Essays.* Vintage, 1970
Darkwood, Vic. *The Lost Art of Travel: A Handbook for the Modern Adventurer.* John Murray, 2006
Davies, Sara. *Global Politics of Health.* Polity Press, 2010
Didion, Joan. *Slouching Towards Bethlehem.* Farrar, Straus and Giroux, 1990
Hibell, Ian. *Into the Remote Places.* HarperCollins, 1985
Humphreys, Alastair. *Thunder and Sunshine.* Eye Books, 2008
Lee, Laurie. *As I Walked Out One Midsummer Morning.* Penguin Books, 1979
Lilwall, Rob. *Cycling Home from Siberia.* Hodder & Stoughton, 2009
Marmot, Michael. *The Health Gap: The Challenge of an Unequal World.* Bloomsbury, 2015

Murphy, Dervla. *Full Tilt: Ireland to India with a Bicycle.* Overlook Press, 1987
O'Hanlon, Redmond. *In Trouble Again: A Journey Between the Orinoco and the Amazon.* Penguin Books, 1989
Onyebuchi Eze, Michael. *Intellectual History in Contemporary South Africa.* Palgrave Macmillan, 2010
Pavitt, Nigel. *Turkana: Kenya's Nomads of the Jade Sea.* Abrams, 1997
Reidy, Eugenie. *Health and Healthcare in Turkana, a Medical Anthropological study for Merlin*, 2010
Sacks, Oliver. *The Man Who Mistook His Wife for a Hat.* Picador, 1986
Silver, George A. 'Virchow, The Heroic Model in Medicine: Health Policy by Accolade'. *American Journal of Public Health*, vol. 77, no. 1, 1987
Speake, Jennifer, ed. *Literature of Travel and Exploration: An Encyclopedia.* Routledge, 2014
Theroux, Paul. *Figures in a Landscape: People and Places.* Hamish Hamilton, 2018

Part Two: Ushuaia to Deadhorse

Glaisher, James. *Travels in the Air.* Richard Bentley & Son, 1871
Macfarlane, Robert. *Mountains of the Mind: A History of a Fascination.* Granta, 2004
MacKinnon, J. B. 'The Problem with Nature Therapy'. *Nautilus Magazine*, 21 Jan 2016
Miller, Henry. *Big Sur and the Oranges of Hieronymus Bosch.* New Directions, 1957
Nabokov, Vladimir. 'The Art of Literature and

Commonsense', in *Lectures on Literature*. Mariner Books, 2002

Proust, Marcel. *In Search of Lost Time*. Modern Library, 2003

Thompson, Hunter S. *The Proud Highway: Saga of a Desperate Southern Gentleman 1955–1967*. Ballantine Books, 1998

Thoreau, Henry David. *Walden*. Princeton University Press, 2004

Part Three: Melbourne to Mumbai

Appiah, Kwame Anthony. *The Lies That Bind: Rethinking Identity*. Profile, 2018

Baedeker's Traveller's Manual of Conversation in Four Languages: English, French, German, Italian. BiblioLife, 2009

Birchmore, Fred A. *Around the World on a Bicycle*. Cucumber Island Storytellers, 1996

Davis, Mike. *Planet of Slums*. Verso, 2006

Dobson, Mary. *Contours of Death and Disease in Early Modern England*. Cambridge University Press, 2003

Filer, Nathan. *The Heartland: Finding and Losing Schizophrenia*. Faber & Faber, 2019

Foucault, Michel. *Madness and Civilization: A History of Insanity in the Age of Reason*. Vintage, 1988

Fraser, John Foster. *Round the World on a Wheel*. Futura Publications, 1989

Gill, A. A. *Here and There: Collected Travel Writing*. Hardie Grant, 2012

Hakim, Adi B. *With Cyclists Around the World*. Roli Books, 2008

Jamieson, Duncan R. *The Self-Propelled Voyager*. Rowman & Littlefield Publishers, 2015
Keeble, T. W. 'A Cure for the Ague: The Contribution of Robert Talbor (1642–81)'. *Journal of the Royal Society of Medicine*, vol. 90, May 1997
Mecredy, R. J, Stoney, G. *The Art and Pastime of Cycling*. Mecredy & Kyle, 1890
Morris, Jan. *A Writer's World: Travels 1950–2000*. Faber & Faber, 2004
Orwell, George. *Down and Out in Paris and London*. Mariner Books, 1972
Reid, Carlton. *Roads Were Not Built for Cars*. Island Press, 2015
Scott, James C. *The Art of Not Being Governed: An Anarchist History of Upland Southeast Asia*. Yale University Press, 2009
Sithirith, Mak, Grundy-Warr, Carl. *Floating Lives of the Tonle Sap*. Regional Center for Social Science, Chiang Mai University, 2013
Tatchell, Frank. *The Happy Traveller: A Book for Poor Men*. Methuen, 1924
Watts, Alan W. *The Book on the Taboo Against Knowing Who You Are*. Vintage, 1989

Part Four: Hong Kong to Calais

Abbey, Edward. *Desert Solitaire*. Ballantine Books, 1985
Bynum, Helen. *Spitting Blood: The History of Tuberculosis*. Oxford University Press, 2012
Dikötter, Frank. *Mao's Great Famine: The History of China's Most Devastating Catastrophe*. Bloomsbury, 2010

Dillard, Annie. *Pilgrim at Tinker Creek*. Harper Perennial, 2000

Dormandy, Thomas. *The White Death: A History of Tuberculosis*. New York University Press, 2002

Fox, Adam T. et al. 'Medical Slang in British Hospitals'. *Ethics & Behavior*, vol. 13, no. 2, 2003, pp. 173–189

Gandy, Matthew. *The Return of the White Plague: Global Poverty and the New Tuberculosis*. Verso, 2003

Harrison, Mark. *Disease and the Modern World: 1500 to the Present Day*. Polity Press, 2004

Hays, J. N. *The Burdens of Disease: Epidemics and Human Response in Western History*. Rutgers University Press, 1998

Hessler, Peter. *River Town: Two Years on the Yangtze*. Harper Perennial, 2006

Hua, Yu. *China in Ten Words*. Pantheon, 2011

Huth, Edward J., Murray, T. J., eds, *Medicine in Quotations: Views of Health and Disease Through the Ages*. The American College of Physicians, 2000

Klein, Naomi. *The Shock Doctrine: The Rise of Disaster Capitalism*. Metropolitan Books, 2007

McMillen, Christian W. *Discovering Tuberculosis: A Global History, 1900 to the Present*. Yale University Press, 2015

McNeill, William H. *Plagues and Peoples*. Anchor, 1977

Miller, Tom. *China's Urban Billion: The Story behind the Biggest Migration in Human History*. Zed Books, 2012

Morrison, Alexander. 'Stalin's Giant Pencil: Debunking a Myth About Central Asia's Borders'. Eurasianet.org, 13 February, 2017

Newby, Eric. *A Short Walk in the Hindu Kush*. Adventure Library, 1999

Nordin, Virginia Davis, Glonti, Georgi. 'Thieves of the Law

and the Rule of Law in Georgia'. *Caucasian Review of International Affairs*, vol. 1, no. 1, 2006

Pisani, Elizabeth. *The Wisdom of Whores: Bureaucrats, Brothels, and the Business of AIDS*. Granta, 2008

Pisani, Elizabeth. 'The Art of Medicine: Tilting at Windmills and the Evidence Base on Injecting Drug Use'. *The Lancet*, vol. 376, 24 July, 2010

Rossabi, Morris. *Modern Mongolia: From Khans to Commissars to Capitalists*. University of California Press, 2005

Sacks, Oliver. *Gratitude*. Knopf Canada, 2015

Said, Edward. *Reflections on Exile and Other Essays*. Harvard University Press, 2002

Solnit, Rebecca. *A Field Guide to Getting Lost*. Penguin Books, 2006

Solnit, Rebecca. *Hope in the Dark*. Canongate Books, 2005

Sontag, Susan. *Illness as Metaphor*. Farrar, Straus and Giroux, 1988

Stevens, Thomas. *Around the World on a Bicycle*. Stackpole Books, 2000

Thomas, Lewis. *The Medusa and the Snail: More Notes of a Biology Watcher*. Penguin Books, 1995

West, Geoffrey. *Scale: The Universal Laws of Life and Death in Organisms, Cities and Companies*. Weidenfeld & Nicolson, 2017

Part Five: Home

Beck, Julie. 'When Nostalgia Was a Disease'. *The Atlantic*, 14 August, 2013

Gawande, Atul. 'Curiosity and What Equality Really Means'. *The New Yorker*, 2 June, 2018

Global Health Watch 5: An Alternative World Health Report. Zed Books, 2017

Kapuscinski, Ryszard. 'Herodotus and the Art of Noticing'. Lettre Ulysses Award Key Note Speech, 2003

Lee, Laurie. *I Can't Stay Long.* Chivers, 1995

Orwell, George. *The Lion and Unicorn: Socialism and the English Genius.* Penguin Books, 1982

Pinker, Steven. *Enlightenment Now: The Case for Reason, Science, Humanism, and Progress.* Viking, 2018

Russell, Bertrand. A *History of Western Philosophy.* Simon & Schuster, 1986

Thomas, Lewis. *The Lives of a Cell: Notes of a Biology Watcher.* Penguin Books, 1978

Acknowledgements

So many people supported, influenced and inspired my journey, or journeys, really, because sitting at a desk and putting down words means casting off and coming home too. David Livingstone once said that it was far easier to travel than to write about it, and he was almost mauled to death by a lion, and spent much of his life with the alternating agonies of malaria and dysentery. I'm just glad I survived.

I must start by thanking my one-woman support team: Mum. And to George Root for your support, and for supporting the support (an expedition, of sorts, too?).

To my editor and friend James Spackman, thank you for your guidance, critical eye and for being such a wise and hard-working advocate for this book. And my thanks to all those hard-working people at Profile, Hannah Ross, Penny Daniel, Susanne Hillen and everyone else who helped summon this book into being.

I'm immensely lucky for the early mentor I found in Tom Griffiths: without your zeal a teenage bike ride through Chile would never have snowballed into this.

I'd like to extend my gratitude and apologies to some fellow riders: Claire, thank you for all the good times. I'm sorry for hogging the map. Olly Davy, thanks for the laughs. I'm sorry for farting into the slipstream. Nyomi Rowsell, thank you for teaching me the art of fuck-this-I'm-doing-it-anyway. I'm sorry for overcooking the noodles. You were all better company than I deserved.

Bikers Nicky Gooch, Nate Roter, Mike Roy, Liyan, James and Sam Lovell all brightened up my days too, but while they appear in the pages of this book, other bikers don't, and I'm sorry I was unable to include you in this story when you were so important to the journey.* I hope you'll understand that one cannot completely reflect the other, given that I was tasked with writing one book about six years on the road. Patagonia was far more fun within the peloton that was Tim Van Meer and Vincent Blusseau. It's impossible to forget towing roller-blader Kay Makishi across Uzbekistan with Stuart Charters and Charles Brands. Extra shout-outs to Aurélie and Layko, James Peacock and Tom Henson-Webb, Martin Lohmann Møller and Susanne van Aardenne, Nick Harman Brown and Romain, and Anaïs and Gilles.

To paraphrase Steve Martin, some people have a way with words, other people, not have way. I'm indebted to all the friends who do have way and could tell me when my choice of words was ungood or rubble. Foremost, my friend Jessica Brook, who I feel very fortunate to have on my side and whose thoughtful feedback was transformative, as I knew it would be. Regular philosophising outside café Oto with Julian Sayarer was also critical to the success of this book, and my own sanity.

I was lucky to receive bits of feedback from friends who were busy with their own writing, among them Henry Wismayer, Elizabeth Gowing, Fearghal O'Nuallain, Jamie Andrews, Tommy Norton, Julian and Olly Davy again.† Others cast an eye on the odd chapter: thanks to Ronan Ryan, Ellie Tait, Jossie Stevens, Zoe Cricks and the nurturing, gloves-off critique of

* In the eyes of my publisher, however, you were worthless bit characters who didn't add to the narrative.

† With respect to any factual errors or poor writing, all of these people shoulder serious responsibility.

The Itinerant Writers Club – Liz Cleere, Helen Moat, Suzy Pope, Paola Fornari, Moira Ashley, Elizabeth and the rest … you're all incredibly bad-ass, though I know you'd prefer a better adjective, and that the superfluous adverb made you cringe inside.

My hosts were numerous and more than hospitable – I'm enormously grateful to those who shined a light on their corner of the world. I have old friends, strangers with impromptu invitations, and countless members of the couchsurfing and warm-showers communities to thank. There are of course too many to mention here, but props must go to these kind souls in particular: Tommy and Roger Bennett, Edit Dékány, Simona Štíchová, Eddie O'Callaghan, Tunc Akdogan, Ümit Orhan, Jocelyn Pihlaja, Nick Cahill, John Mooney, Ant and Julia Rosevear, Paul Lange, Sean and Jill Ingram, Sugnet and Pierre Smit, Sarah O'Brien, Matt Moran, Ana Maria Giraldo, Fin O'Hara and John Onken, Rachel Thomas, Alan O'Hara and Ino O'Hara Valerio, Ryan Flegal, Benny and Jo Merritt-Hall, Cat and Pat Patterson, Becky Krenz, Elizabeth Kroupa, Tammy, Kait and Dave Kanaris, MaryLouise O'Driscoll and Paul Farnan, Duncan Edwards, Sage Cohen, Ian and Sarah Humble, Eddie FitzPatrick, Peter and Dermot O'Driscoll, Dion Peter and Pune Thomas, Simon McCrum, Anne Suchanecki, Philippe and Lewis Danielski, Andrew X. Pham, Ian Ferguson, Leemax and Pedal Attack, Ankan Dé, Rahul Antao and Annelie Bernhart, Elizabeth Hacker and Sanju, Anna Chiumento, Oskar Ruch, Rob and Christine Lilwall, Rachael Desgouttes, Swee kian Tan, Andy Peat and Wai Yeng Chan, Medina, Wahed Bazargan, Froit VanderHarst and Paul Byler.

Thank you as well to all the people behind medical projects that allowed me to visit and were kind enough to explain. Again, only some have been included in this book. Though I have not changed details, I have sometimes changed the names of patients and clinicians in the text to preserve confidentiality, but thank you. Additionally, I have the following people I can name to thank: Dr Umed Nazrishoev, Nizoramo Ramikhudoeva and Dr Mahbut Bahromov from The Aga Khan Foundation, Jon Morgan, Binaifer Jesia, Dr Irina Amongpradja and Sekolah Kami, Professor François Nosten at Shoklo Malaria Research Unit, The Children Welfare Centre in Kathmandu, Anandaban Leprosy Hospital, Dr Rahimullah Hamid, Pink Armenia, Knarik Khudoyan, Vladimir Lozinski, The Roddy Scott Foundation, Cathy McLain and the McLain Association for Children, Gayle Nixon, Pippa Findlay and all the staff who worked for Merlin.

Other people had a hand in engineering and steering this journey, lending expert advice, sponsorship, skills, connections, inspiration and donations. Thank you to Alasdair and Shelagh Scadding at MSG Bikes, Mariano Barnes for help with images, Sean Lally at Cycle Systems Academy, Will Kerley, Dr Nicholas Hart, Dylan Reynolds at Ride and Seek, Lou Hamilton, Oliver Thomas, Rayhan Demytrie at the BBC, Howard Carter, Michael Urquhart, David Picatti, David Johnstone, Thomas Carley, Neil and Harriet Pike for the trailblazing andesbybike.com, Alistair Humphreys, Professor Alan Fenwick, Simon Hughes MP, the Alaskan World Affairs Council, Abingdon School, Nurofen Big Lives Trust, Lyon Equipment, Buff, Madison, Burton-McCall, Endura, Schwalbe, Hilleberg, Brooks, Shane Winsor at the Royal Geographical Society, Tom Wingfield, Nik Voigt and Mariam Chachia and all the staff at St Thomas' A&E department. Honestly, I could go on, but there is also my editorially directed word limit to consider and I've had to cut a sparkling metaphor about an Indonesian sunrise to pay tribute to family and friends: Ronan Ryan, Siobhan Ryan, Angela Briggs, Roisin O'Hara, Françoise Holland and John Fabes.

And finally, to all the strangers. To the people who took me in, kept me warm, prepared food for me and politely ignored the speed and grotesquery of my eating. Thank you. You fuelled my hope, too.

Author's Note

In medicine, it is deemed important to protect the privacy and confidentiality of patients. I would like to reassure everyone, but especially the General Medicine Council – who have the power to revoke my medical licence – that I, just like Hippocrates, think that this is very important too. To this end, I have occasionally altered names, places and minor details of events centred on healthcare.